SAUDI ARABIA: TERRORISM, U.S. RELATIONS AND OIL

AF848923

SAUDI ARABIA: TERRORISM, U.S. RELATIONS AND OIL

NINO P. TOLLITZ
EDITOR

Nova Science Publishers, Inc.
New York

Copyright © 2005 by Nova Science Publishers, Inc.

All rights reserved. No part of this book may be reproduced, stored in a retrieval system or transmitted in any form or by any means: electronic, electrostatic, magnetic, tape, mechanical photocopying, recording or otherwise without the written permission of the Publisher.

For permission to use material from this book please contact us:
Telephone 631-231-7269; Fax 631-231-8175
Web Site: http://www.novapublishers.com

NOTICE TO THE READER

The Publisher has taken reasonable care in the preparation of this book, but makes no expressed or implied warranty of any kind and assumes no responsibility for any errors or omissions. No liability is assumed for incidental or consequential damages in connection with or arising out of information contained in this book. The Publisher shall not be liable for any special, consequential, or exemplary damages resulting, in whole or in part, from the readers' use of, or reliance upon, this material.

This publication is designed to provide accurate and authoritative information with regard to the subject matter covered herein. It is sold with the clear understanding that the Publisher is not engaged in rendering legal or any other professional services. If legal or any other expert assistance is required, the services of a competent person should be sought. FROM A DECLARATION OF PARTICIPANTS JOINTLY ADOPTED BY A COMMITTEE OF THE AMERICAN BAR ASSOCIATION AND A COMMITTEE OF PUBLISHERS.

LIBRARY OF CONGRESS CATALOGING-IN-PUBLICATION DATA
Available upon request

ISBN 1-59454-729-7

Published by Nova Science Publishers, Inc. ✣ New York

CONTENTS

PREFACE

Saudi Arabia, a monarchy ruled by the Saud dynasty, enjoys special importance in much of the international community because of its unique association with the Islamic religion and its oil wealth. Since the establishment of the modern Saudi kingdom in 1932, it has benefitted from a stable political system based on a smooth process of succession to the throne and a prosperous economy dominated by the oil sector. With one-fourth of the world's proven oil reserves and some of the lowest production costs, Saudi Arabia is likely to remain the world's largest net oil exporter for the foreseeable future. During January-October 2004, Saudi Arabia supplied the United States with 1.5 million barrels per day of crude oil, or 15%, of U.S. crude oil imports during that period.The September 11, 2001 attacks fueled criticisms within the United States of alleged Saudi involvement in terrorism or of Saudi laxity in acting against terrorist groups. Of particular concern have been reports that funds may be flowing from Saudi Arabia and other Middle East countries to terrorist groups, largely under the guise of charitable contributions. Critics of Saudi policies have cited a number of reports that the Saudi government has permitted or encouraged fund raising in Saudi Arabia by charitable Islamic groups and foundations linked to Osama bin Laden's Al Qaeda organization or like-minded entities.

In: Saudi Arabia: Terrorism, US Relations and Oil ISBN 1-59454-729-7
Editor: Nino P. Tollitz, pp. 1-26 © 2005 Nova Science Publishers, Inc.

Chapter 1

SAUDI ARABIA: CURRENT ISSUES AND U.S. RELATIONS*

Alfred B. Prados

SUMMARY

Saudi Arabia, a monarchy ruled by the Saud dynasty, enjoys special importance in much of the international community because of its unique association with the Islamic religion and its oil wealth. Since the establishment of the modern Saudi kingdom in 1932, it has benefited from a stable political system based on a smooth process of succession to the throne and a prosperous economy dominated by the oil sector. Some commentators have suggested that growing internal pressures and regional tensions may have weakened the Saudi regime in recent years; others point to an apparent consensus behind Saudi institutions over much of the last century as evidence of long-term stability.

The United States and Saudi Arabia have long-standing economic and defense ties. A series of informal agreements, statements by successive U.S. administrations, and military deployments have demonstrated a strong U.S. security commitment to Saudi Arabia. Saudi Arabia was a key member of the allied coalition that expelled Iraqi forces from Kuwait in 1991. Saudi Arabia subsequently hosted U.S. aircraft enforcing the no-fly zone over southern Iraq. Saudi officials expressed opposition to the U.S.-led military campaign launched against Iraq in March 2003 (Operation Iraqi Freedom), although Saudi Arabia

* Excerpted from CRS Report IB93113, dated February 4, 2005.

reportedly permitted certain support operations by U.S. and British military forces, in addition to making some facilities available to them. By mutual agreement, the United States withdrew virtually all its forces from Saudi Arabia at the end of August 2003.

Bombing attacks against several U.S. and foreign operated installations in Saudi Arabia have raised some concerns about security of U.S. personnel and what appears to be growing anti-Americanism in some segments of the Saudi population. Since the attacks on the United States September 11, 2001, some commentators have maintained that Saudi domestic and foreign policies have created a climate that may have contributed to terrorist acts by Islamic radicals. U.S. officials have generally cited Saudi support in the aftermath of the attacks, including increased intelligence sharing, law enforcement activities, and tracking of terrorist financing.

In its final report, released on July 23, 2004, the U.S. National Commission on Terrorist Attacks upon the United States (the 9/11 Commission) described Saudi Arabia as having been "a problematic ally in combating Islamic extremism," while noting that Saudi cooperation has improved, especially since further terrorist attacks in Saudi Arabia beginning in May 2003. The National Intelligence Reform Act (P.L. 108-458, December 17, 2004) contains a requirement (Section 7119(b)(3)) that the President submit to designated congressional committees a strategy for collaboration with Saudi Arabia, as part of a larger report on U.S. government activities to implement the provisions of this act.

Other principal issues of bilateral interest include security in the post-war Gulf region, the Saudi position on the Arab-Israeli conflict, arms transfers to Saudi Arabia, Saudi external aid programs, bilateral trade relationships, and Saudi policies involving human rights and democracy. The Consolidated Appropriations Act for FY2005 (P.L. 108-447, December 8, 2004) contains a ban on U.S. assistance to Saudi Arabia but provides for a presidential waiver (Section 575) if the President certifies that Saudi Arabia is cooperating in the war against terrorism.

Most Recent Developments

Section 7119(b)(3) of S. 2845, the National Intelligence Reform Act of 2004 (also known as the 9/11 Commission Implementation Act of 2004) requires the President to submit to designated congressional committees a strategy for collaboration with Saudi Arabia, as part of a larger report on U.S. government activities to implement the provisions of this act. President Bush signed S. 2845 on December 17, 2004, as P.L. 108-458. Also, Section

575 of the Consolidated Appropriations Act for FY2005 (P.L. 108-447, December 7, 2004) bans U.S. assistance to Saudi Arabia but provides waiver authority if the President certifies that Saudi Arabia is cooperating in the war against terrorism.

According to press reports, the Saudi Minister of Defense and 2nd Deputy Prime Minister, Prince Sultan bin Abd al-Aziz, told Saudi newspapers on January 25, 2005, that the country's appointive consultative ("Shura") council will be increased from 120 to 150 members within the next three months and that its authority will be expanded. Meanwhile, campaigning has begun for municipal council elections scheduled to begin on February 10 as part of a three-stage process.

Background and Analysis

Current Issues

Oil and national security concerns have combined to produce a close and cooperative relationship between the United States and Saudi Arabia for much of the past century. Since the award of the first Saudi oil concession to a U.S. company in 1933, both states have had an increasing interest, respectively, in the marketing and acquisition of Saudi petroleum supplies. As regional threats multiplied in the latter half of the century, mutual concerns over the stability of Saudi Arabia and other moderate regimes in the Arabian Peninsula engendered a significant degree of defense cooperation. Saudi Arabia was a key member of the allied coalition that expelled Iraqi forces from Kuwait in 1991, and for the next 12 years it hosted more than 5,000 U.S. military personnel, most of them from U.S. Air Force units that enforced a no-fly zone over southern Iraq (Operation Southern Watch). The U.S. military presence, however, aroused resentment among many Saudis, particularly those inclined toward Islamic fundamentalism, and engendered both domestic and regional criticism of the Saudi leadership for allowing non-Muslim soldiers to be stationed in Saudi Arabia, where the two holiest sites in the Islamic religion are located. In August 2003, by mutual U.S.-Saudi agreement, almost all U.S. military personnel were withdrawn from Saudi Arabia, but the two countries will continue to conduct joint training activities and military exercises. (See below, *Future U.S. Military Presence*.)

	Saudi Arabia in Brief
Population (July 2004):	25,795,938 (includes 5,576,076 foreign residents) Growth rate: 2.44%
Area:	1,960,582 sq.km. (756,985 sq.mi.); just over one fifth the size of the United States
Ethnic Groups:	(native Saudis only) Arab 90%; Afro-Asian 10%
Religion:	(native Saudis only) Muslim 100% (Sunni 85-95%, Shi'ite 5-15%)
Literacy (2003):	78.8% (male 84.7%, female 70.8%)
GDP (2003):	$211.6 billion; growth rate: 6.4%*
External Public Debt (2003):	$39.1 billion
Inflation (2003):	1.0%
Unemployment (2003):	25%

*Unusually high figure, largely owing to surge in oil production.

Sources: IMF; U.S. Dept. of Commerce; CIA World Factbook; Economist Intelligence Unit (London)

September 11 Terrorist Attacks and Aftermath

Allegations

The September 11, 2001 attacks kindled criticisms within the United States of alleged Saudi involvement in terrorism or of Saudi laxity in acting against terrorist groups. Commentators have pointed to the high percentage of Saudi nationals among the hijackers (15 out of 19). Some critics go so far as to accuse Saudi government officials of responsibility for the September 11, 2001, attacks through design or negligence and for the continuing threat posed by the perpetrators or by like-minded terrorist groups. Others maintain that Saudi domestic and foreign policies have created a climate that may have contributed to terrorist acts by Islamic radicals. For example, some believe that the Saudi regime has fostered international terrorism by funding religious schools (*madrasas*) that propagate extreme forms of Islam and advocate violence. (For more information on these schools, see CRS Report RS21654, *Islamic Religious Schools, Madrasas: Background*, by Febe Armanios.) Critics of Saudi policies have also cited a multiplicity of reports that the Saudi government has permitted or encouraged fund raising in Saudi Arabia by charitable Islamic groups and foundations linked to Osama bin Laden's Al Qaeda organization, which the U.S. government has identified as clearly responsible for the hijacks. The exiled bin Laden was formerly a Saudi national, but Saudi authorities revoked his citizenship in 1994.

Saudi Responses

Saudi officials maintain that they are working closely with the United States to combat terrorism, which they say is aimed as much at the Saudi regime as it is at the United States. In October 2001, the Saudi government announced that it would implement U.N. Security Council Resolution 1373, which called among other things for freezing terrorist related funds. The Saudi government later invited the G-7 Financial Action Task Force (FATF) on Money Laundering into the kingdom to conduct a "mutual evaluation," established a Financial Intelligence Unit (FIU) to collect intelligence on financial activity, and passed new banking regulations.[1] On February 24, 2004, Saudi Arabia and 12 other Middle East or Asian countries attended a conference organized by the 29-member FATF, where attendees focused on terrorist use of charitable organizations to finance terrorist operations.

Saudi officials say they have arrested suspected terrorists including persons believed to be affiliated with Al Qaeda and sought to dampen religious extremism in the Kingdom. In a speech to an Islamic conference on August 30, 2003, for example, Saudi King Fahd bin Abd al-Aziz condemned terrorism and exhorted Muslim clerics to emphasize peace, security, cooperation, justice, and tolerance in their sermons. Press reports indicate that Saudi security forces had arrested 600 suspects as of November 10, 2003. On several occasions, most recently in September 2003, the Saudi Embassy issued a comprehensive white paper detailing initiatives undertaken by Saudi Arabia in the war against terrorism since the September 11, 2001 attacks, with particular emphasis on apprehension of suspected terrorists, establishment of joint task forces with the United States (see below), intelligence cooperation, and various steps against terrorism financing.

U.S. government statements have generally complimented Saudi cooperation with the U.S. campaign against terrorism, while sometimes suggesting that the Saudi government could do more. In its most recent annual report entitled *Patterns of Global Terrorism, 2003* (published April 29, 2004), the State Department noted that the Saudi government adopted a new law in August 2003 criminalizing money laundering and terrorist financing, and also established new financial controls to regulate charitable contributions. In this connection, according to the U.S. Treasury Department, the United States and Saudi Arabia have worked together in jointly designating several entities as supporters of terrorism. For example, on June 2, 2004, a spokesman of the Saudi Embassy in Washington, D.C. announced that a large Saudi-based charitable organization (the al-Haramayn Islamic Foundation), which has been linked to terrorist activity, would be dissolved and that its assets, along with the assets of other Saudi charitable organizations, would be merged into a new organization to be called the Saudi National Commission for Relief and Charity Work Abroad, in an effort "to ensure that the charity of our citizens goes to those who need it." During the same period, the United States and Saudi Arabia reportedly asked the U.N. Sanctions Committee to add five other overseas branches of al-Haramayn to its list of foreign terrorist organizations.

Joint Congressional Report

On July 24, 2003, the House and Senate Intelligence Committees released part of a 900-page report entitled Joint Inquiry into Intelligence Community Actions before and after the Terrorist Attacks of September 11, 2001 (S.Rept. 107-351; H.Rept. 107-792). The Bush Administration refused to allow the release of an approximately 28-page section of the report. According to press articles, persons who claim to have read the still-classified section of the report say it covers Saudi links with individuals involved in the September 11 attacks; specifically, the classified section reportedly states that senior Saudi officials channeled hundreds of millions of dollars to charitable groups that may have helped fund the attacks.[2] Saudi officials, including the Saudi Ambassador to the United States, have denounced the report, maintaining that "Al Qaeda is a cult seeking to destroy Saudi Arabia as well as the United States. By what logic would we support a cult that is trying to kill us?"[3] On July 29, in response to an urgent request from Saudi Arabia, President Bush met with Saudi Foreign Minister Saud al-Faysal, who called for the release of the still classified section of the report to enable Saudi Arabia to rebut the allegations contained therein. President Bush refused to do so on the grounds that disclosure could reveal intelligence sources and methods to enemies of the United States and might compromise the on-going investigation of the September 11, 2001, attacks.[4]

Joint Task Force

In August 2003, the United States and Saudi Arabia reached an agreement for the establishment in Saudi Arabia of a joint U.S.-Saudi task force to monitor terrorist financing. The task force, composed of representatives from the U.S. Treasury Department, the FBI, and other U.S. agencies, will screen bank accounts, computer records, and other financial data in an effort to track the flow of money to terrorist organizations and shut it off. U.S. officials left for the Saudi capital of Riyadh on August 26, to work out details regarding the task force. News articles have noted that this is the first time Saudi authorities have allowed U.S. government agencies to maintain an extended presence in Saudi Arabia or provided them with access to Saudi documents and investigations. Some speculate that the May 2003 bombings of three residential compounds in Riyadh (see below) in which 34 people died, including eight Americans and seven Saudis, may have impelled Saudi authorities to expand cooperation with the United States in fighting terrorism ("U.S.- Saudi Anti-Terror Operation Planned," *Washington Post*, August 26, 2003).

9/11 Commission Report

In its final report, released on July 23, 2004, the U.S. National Commission on Terrorist Attacks upon the United States (the 9/11 Commission) described Saudi Arabia as having been "a problematic ally in combating Islamic extremism." The report takes note of long-standing cooperative relations between the U.S. and Saudi governments, growing misunderstandings at the popular level in recent years, and U.S. criticisms in the aftermath of the 9/11 attacks that Saudi officials could do more to fight terrorism. The report acknowledges increased efforts in that regard since mid-2003 when terrorists began hitting targets in Saudi Arabia itself with more frequency; today, according to the report, "The Kingdom of Saudi Arabia is now locked in mortal combat with al Qaeda." One of the key recommendations in the 9/11 report addresses the U.S.-Saudi relationship:

> The problems in the U.S.-Saudi relationship must be confronted, openly. The United States and Saudi Arabia must determine if they can build a relationship that political leaders on both sides are prepared to publicly defend — a relationship about more than oil. It should include a shared commitment to political and economic reform, as Saudis make common cause with the outside world. It should include a shared interest in greater tolerance and cultural respect, translating into a commitment to fight the violent extremists who foment hatred.

Major Terrorist Attacks

The mid-1990s saw two forerunners of more recent attacks by terrorists against U.S. and other targets starting in 2003. The first attack, which occurred on November 13, 1995, at the headquarters of a U.S. training program for the Saudi National Guard in the capital of Riyadh, killed seven persons (including five U.S. citizens). Several months later, Saudi authorities executed four Saudis, who reportedly confessed to having been influenced by Islamic fundamentalist exiles. The second and more lethal explosion occurred in June 1996 at Khobar Towers, a housing facility for U.S. Air Force personnel near Dhahran Air Base, killing 19 U.S. Air Force personnel and wounding many others. Saudi Interior Minister Prince Nayif told reporters on May 22, 1998, that the bombings "were carried out by Saudis with the support of others." On June 21, 2001, U.S. Attorney General John Ashcroft announced that a federal grand jury had indicted 14 individuals in connection with the explosion, which killed 19 U.S. servicemen and wounded 372 others. According to the U.S. Justice Department, the suspects belonged to the Saudi and Lebanese branches of the militant Shi'ite Muslim organization Hizballah, which is supported by Iran. The 1996 Khobar bombing remains unresolved, but seven years passed before another incident of this nature took place.

Riyadh, May 2003

Three near-simultaneous suicide bombings at about 11:20 p.m. on May 12, 2003, destroyed three housing compounds in Riyadh, Saudi Arabia, inhabited by U.S. military personnel and other foreign residents in Riyadh. According to reports, as many as 34 people were killed, including eight Americans, seven Saudis, two Jordanians, two Filipinos, one Lebanese, one Swiss and nine unidentified persons (possibly including the perpetrators), while many more were wounded. A fourth explosion occurred the following morning at the headquarters of a U.S.-Saudi joint venture, but there were no reported injuries. U.S. and Saudi officials said the attacks had many of the earmarks of previous Al Qaeda operations in their coordinated timing and in their techniques, in that the assailants attacked guards at the compound gates and then drove their trucks into the compounds. President Bush condemned the attack, which he linked to Al Qaeda.[5] Top Saudi leaders, including Crown Prince Abdullah and 2nd Defense Minister Prince Sultan, also condemned the attacks, and some commentators expressed the view that the attacks were targeted against the Saudi government as well. On May 13, a U.S. interagency team including six FBI representatives left for Saudi Arabia to assist in investigating the attack. A joint U.S.-Saudi task force was established several months later, as discussed above.

Riyadh, November 2003

On November 9, a car bomb exploded in the Saudi capital of Riyadh at a residential compound inhabited mainly by foreign residents from other Arab countries, killing at least 17 and wounding over 120, including four Americans. According to a statement by the Saudi Cabinet, King Fahd promised to "strike with an iron fist" at anyone seeking to threaten security and stability in the country; Interior Minister Prince Nayif said there would be no dialog with the perpetrators "other than with the rifle and the sword"; and the Minister of Islamic Affairs described the attack as "flagrant aggression" against Islam. U.S. and Saudi officials blamed Al Qaeda for the attack; visiting U.S. Deputy Secretary of State Richard Armitage said "this attack bears their hallmark." Noting that virtually all U.S. military personnel have departed the kingdom, some commentators suggested that the perpetrators were primarily seeking to destabilize the Saudi government.

Riyadh, April 2004

On April 21, 2004, a suicide bomber detonated a Chevrolet Blazer packed with explosives in front of the Saudi General Security building in the capital of Riyadh, killing at least four persons (two Saudi police officers, one civilian, and one Syrian girl) and wounding 148. Al-Haramayn Brigades, a group allegedly connected to Al Qaeda, claimed responsibility for the attack, which was the first suicide bombing of a major Saudi government building. Saudi officials said the perpetrators "are not what they claim to be: people of Islam" but "criminals."

Yanbu, May 2004

During a shooting spree on May 2, 2004, four gunmen killed two U.S., one Australian, and two British citizens employed by the Swiss-Swedish based engineering company ABB Lummus, which was engaged in upgrading operations at a petrochemical plant in the Saudi Red Sea port city of Yanbu. The plant is jointly owned by U.S. Exxon Mobil and a Saudi industrial firm. The gunmen dragged the body of one U.S. victim from the back of a car and fired at a nearby McDonald's and a Holiday Inn before being shot by members of the Saudi Arabian National Guard. Two Saudi security officers were killed while trying to apprehend the gunmen. No group has claimed responsibility, but Saudi Interior Minister Prince Nayif blamed Al Qaeda for being behind the shooting, and a statement of undetermined authenticity by Abdul Aziz al-Muqrin, the leader of an apparent Al-Qaeda affiliate known as "Al-Qaeda in the Arabian Peninsula," on May 5 praised "the four heros" who carried out the attack. Meanwhile, Foreign Minister Prince Saud al-Faysal accused a London-based dissident group, the Movement for Islamic Reform in Arabia. In addition, Crown Prince Abdullah and Prince Saud suggested that "Zionists" had manipulated some of the terrorists but gave no evidence to support this claim.

Khobar, May 2004

On May 29, 2004, at least four gunmen attacked a residential compound housing oil company employees and dependents in Khobar, Saudi Arabia, seizing approximately 50 hostages. Twenty-two of the hostages, including one U.S. citizen, were killed and approximately 25 wounded before the hostage-takers fled during a rescue operation mounted by Saudi commandos on May 30. Three gunmen either escaped or were allowed to leave by Saudi authorities in return for releasing the hostages; a fourth, described as a ringleader, was wounded and captured. In a taped statement, Al Qaeda operative Abd al-Aziz al-Muqrin (see above) claimed responsibility for the attack and promised that the year 2004 would be "bloody and miserable for infidels."

Assassinations and Murders, Post-May 2004

Since 2004, terrorist groups in Saudi Arabia have targeted several western expatriates, including three U.S. contractors working on military-related projects. This shift by terrorists from bombs to targeted assassinations has been described by some observers as an attempt by the perpetrators to deflect mounting criticism of Saudi militants for previous indiscriminate bombings that cause injury and death to increasing numbers of Saudis and other Muslims.[6] On June 8 and 12, respectively, two U.S. contractors were shot and killed near their residences. On June 12, another U.S. contractor, Lockheed Martin employee Paul M. Johnson, was kidnapped by Muqrin's group, which subsequently gave Saudi authorities a three-day deadline to release detained members of a militant Islamist cell. On June 18, Muqrin's "Al-Qaeda in the Arabian Peninsula" posted photos in which they claimed to have beheaded Johnson after Saudi authorities did not meet the three-day deadline. Meanwhile, an estimated 15,000 members of Saudi security forces mounted a massive hunt to find Johnson and his captors, and on June 19, Saudi officials announced that they had killed Muqrin and three of his confederates in a shootout in the capital of Riyadh. In a CNN interview on June 21, a senior Saudi foreign policy advisor described the elimination of Muqrin as a major blow to his group but added that others might seek to reconstitute the group's capability. Subsequently, conflicting press reports indicated that Muqrin's replacement, Saleh Al-Oufi, was killed by Saudi security forces and replaced by Saud Al-Otaibi, but a Saudi Ministry of Interior officials denied this report.

Jeddah (U.S. Consulate), December 2004

On December 6, 2004, militants using small arms and grenades stormed the heavily fortified U.S. Consulate in the Red Sea port city of Jeddah, killing five foreign employees of the consulate before Saudi security personnel killed four of the militants and captured the other one. The Saudi cabinet condemned the attack, which Saudi officials blamed on "deviants," a term used by Saudi officials to describe operatives or supporters of Al Qaeda. This is the first major terrorist assault inside Saudi Arabia since the attack on the oil company residential compounds at Khobar on May 29. Some commentators speculate that Al Qaeda may have shifted its strategy again to target U.S. and western interests while refraining from attacks on domestic targets, in an effort to recoup Saudi domestic support for their operations.[7]

Evolution of Saudi Efforts

In the past, U.S. officials have criticized Saudi counterparts for insufficient sharing of information that Saudi officials have gained from their investigations of terrorist acts that have killed or injured U.S. citizens. Press reports indicate that U.S.-Saudi cooperation in the investigation of terrorist incidents has improved since mid-2003. Both U.S. and Saudi officials said the impetus for closer cooperation came from the May attacks, which one knowledgeable observer described as "the inevitable wake up call" for Saudi leaders increasingly concerned over apparent attempts by terrorists to target the Saudi regime. The November bombing, which occurred after virtually all U.S. forces had left the country, may have reinforced Saudi concerns over their vulnerability to such attacks by Al Qaeda and like-minded groups,[8] and Saudi willingness to share information with U.S. officials. According to the 9/11 Commission's report, "[a]s in Pakistan, Yemen, and other countries, [Saudi] attitudes changed when the terrorism came home." In another vein, on June 23, 2004, Saudi Crown Prince Abdullah offered a limited amnesty to members of a "misguided group" who turn themselves in to Saudi authorities within one month. According to press reports, only six responded to the offer, one of whom was on the kingdom's 26-member most-wanted list. Meanwhile, in early June, Saudi Ambassador to the United States Prince Bandar bin Sultan promised to conduct war against the insurgents, whom he described as "deviants" and "traitors to their faith" and added: "War means war. It is not Boy Scout camp."[9]

Bin Laden Tape

On December 16, 2004, an audiotape circulated on the internet and purportedly from Osama bin Laden called for an uprising against the Saudi monarchy. The tape charged the Saudi royal family with misrule, condemned other "tyrannical ruling families" in the region, praised the December 6 attack on the U.S. Consulate in Jeddah, exhorted listeners to attack Gulf oil supplies, and added that "Targeting America in Iraq in terms of economy and loss of life is a golden and unique opportunity." On the same date, the London-based Saudi dissident group, "Movement for Islamic Reform in Arabia," called for protesters to mount anti-regime demonstrations in Saudi cities, but these demonstrations failed to materialize.

Saudi Stance on Iraq

Operation Iraqi Freedom

Between the Gulf War of 1991 and Operation Iraqi Freedom in 2003, Saudi Arabia hosted U.S. Air Force units that conducted over flights to enforce a no-fly zone over southern Iraq (Operation Southern Watch). Although they did not usually object to small scale U.S. responses to Iraqi aircraft or air defense units challenging allied aircraft conducting these over flights, Saudi authorities were opposed to large-scale allied military action against Iraqi targets. Saudi Arabia opposed the U.S.-led Operation Iraqi Freedom, and on March 19, 2003 (the day President Bush initiated the campaign), a communiqué by King Fahd stated that Saudi Arabia "will not participate in any way" in the war. A number of news reports, however, indicated that Saudi Arabia informally agreed to provide logistical support to U.S.-led forces: permission to conduct refueling, reconnaissance, surveillance, and transport missions from bases in Saudi Arabia; landing and over flight clearances; and use of a U.S.-built facility in Saudi Arabia known as the Combat Air Operations Center (CAOC) to coordinate military operations in the region.[10] Also, on March 8, 2003, Saudi Defense Minister Prince Sultan bin Abd al-Aziz said his government was allowing U.S. troops to use two airports in northern Saudi Arabia for "help in a technical matter." A later report in the *Philadelphia Inquirer* on April 26, 2004, quoting unnamed U.S. and Saudi officials, alleged that Saudi Arabia had a wider role in the war than had been publicized at the time. In addition to support noted above, the officials said the Saudi royal family permitted the staging of special forces operations from inside Saudi Arabia, allowed some 250-300 mainly transport and surveillance planes to fly missions from

Saudi Arabia, and provided tens of millions of dollars in discounted oil, gas, and fuel for U.S. forces.

Meanwhile, on March 5, 2003, Saudi Arabia dispatched over 1,000 troops to Kuwait, where they formed part of a joint umbrella force from several Gulf countries to help protect Kuwait but did not have an offensive role against Iraq. According to the Saudi Embassy in Washington, Saudi Arabia contributed a total of 6,000 troops to the allied force in Kuwait (known as Peninsula Shield).

Post-War Iraq: Withdrawal of U.S. Troops From Saudi Arabia

Following the collapse of the Saddam Hussein regime, the *New York Times* reported on April 30, 2003, that the United States planned to withdraw almost all of its 5,000 troops in Saudi Arabia and move its Combat Air Operations Center to neighboring Qatar. The U.S. Air Force unit to which most U.S. military personnel in Saudi Arabia had been assigned was formally deactivated on August 27, 2003. On September 22, 2003, the *New York Times* reported that the last American combat troops had left Saudi Arabia earlier in the month. Approximately 220 U.S. military and contract advisors, most of them working with the Saudi National Guard, remain in country. Both U.S. and Saudi officials hoped the departure of U.S. troops would remove some of the pressure on the Saudi government from militant anti-U.S. groups; but some terrorist attacks against residential compounds have continued, as noted above.

Infiltrators?

In late August 2003, a senior State Department official commented that pro-Saddam Arab volunteer fighters have been infiltrating into Iraq through Iran, Syria, and Saudi Arabia to mount attacks against U.S.-led coalition forces in Iraq. In remarks published on August 26, a senior Saudi foreign policy advisor responded that "we have no evidence of Saudis crossing into Iraq and we have received no evidence from the U.S. government."[11] A press article on November 19, 2003, quoted local U.S. military commanders as saying that their soldiers had encountered relatively few foreign fighters crossing into Iraq from neighboring countries, including Saudi Arabia.[12] According to a *New York Times* report of April 23, 2004, quoting Saudi officials, the Saudi government has installed heat sensors to detect movement on the Saudi-Iraqi border in an effort to seal it. In late December 2004, unidentified western diplomats in Saudi Arabia reportedly said that several hundred Saudi nationals are fighting in Iraq from a total of 1,000-1,500 foreign insurgents, but went on to say that most Saudi infiltrators had come via Syria or other countries rather than directly from Saudi Arabia, which has tighter border controls. (London Financial Times, December 20, 2004).

Iraqi Debt

As of January 2004, Iraq reportedly owed the Saudi government $9 billion in debts incurred during the Saddam Hussein regime, while private Saudi firms and banks hold about $19 billion in Iraqi debt.[13]

Arab-Israeli Conflict

Saudi Arabia supports Palestinian national aspirations, strongly endorses Muslim claims in the old city of Jerusalem, and has been increasingly critical of Israel since the outbreak of the Palestinian uprising in the occupied West Bank and Gaza in September 2000. At the same time, Saudi Arabia has supported U.S. policy by endorsing Israeli-Palestinian peace agreements, joined with neighboring Gulf states in 1994 in terminating enforcement of the so-called secondary and tertiary (indirect) boycotts of Israel while retaining the primary (direct) boycott, and adopted a more pro-active approach to peacemaking. In March 2002, Crown Prince Abdullah proposed a peace initiative calling for full Israeli withdrawal from occupied territories in return for full normalization of relations between Arab states and Israel. The plan was endorsed by the Arab League at a summit conference on March 27-28, 2004 and used as a basis of discussion between Crown Prince Abdullah and President Bush at a bilateral meeting in April 2002. Over a year later, on June 3, 2003, President Bush, Crown Prince Abdullah, and four other Arab leaders met at Sharm al-Shaykh, Egypt, where the attendees endorsed the Road Map — a plan for Palestinian-Israeli peace promulgated by the United States, the United Nations, Russia, and the European Union (the "Quartet").

Saudi-Palestinian Relations

Saudi Arabia, like other Arab states, recognizes the Palestine Liberation Organization (PLO) as the legitimate representative of the Palestinian people. Saudi officials say that their support to Palestinians (estimated at $80 million to $100 million per year) is provided exclusively to the Palestinian Authority (PA), which was established under the Israeli-Palestinian agreement of September 13, 1993, known as the first Oslo Accord.[14] Saudi Arabia has provided aid (variously estimated at $33 million and $59 million) to families of Palestinians killed or injured in the three-year-old Palestinian uprising; in addition, Saudis raised additional funds (over $100 million according to one report) for this purpose at a telethon sponsored by King Fahd on April 11, 2002. Saudi officials told U.S. counterparts in late April 2002 that proceeds of the telethon are funneled through non-governmental organizations to provide some humanitarian support to needy Palestinian families; the Saudis drew a distinction between their fund raising activities and those of Iraq under Saddam Hussein, which paid families who would sacrifice their children as suicide bombers. (For more information on Saudi payments to families of Palestinians killed in the Palestinian uprising or imprisoned by Israeli authorities, see CRS Report RL32499, *Saudi Arabia: Terrorist Financing Issues.*)

Hamas

There have been unsubstantiated reports of Saudi assistance to the PLO's principal rival organization, the fundamentalist Hamas organization, which the U.S. government has designated as a foreign terrorist organization. In its annual report on terrorism for the year 2001 (Patterns of Global Terrorism, 2001, published May 21, 2002), the State Department noted that Hamas receives funding from "private benefactors in Saudi Arabia and other moderate Arab states." In its most recent issue (Patterns of Global Terrorism, 2003, published April 29, 2004), the State Department does not specifically mention Saudi Arabia; rather, it states that Hamas "[r]eceives some funding from Iran but primarily relies on donations from Palestinian expatriates around the world and private benefactors, particularly in Western Europe, North America, and the Persian Gulf region." The State Department reports do not estimate amounts involved. According to one press report, people in Saudi Arabia contribute approximately $5 million to Hamas per year, or approximately half of its annual operating budget.[15] Some reports indicate that Saudi authorities tolerate fund raising for Hamas. For example, in May 2002, Israeli officials, citing captured Palestinian documents, said the Saudi government had given money to 13 charities, seven of which provide support to Hamas. Saudi spokesman Adel al-Jubeir, an advisor to Crown Prince Abdullah, maintains that "no Saudi government money goes to Hamas, directly or indirectly."[16]

Arms Transfers to Saudi Arabia

U.S. Arms Sales

The United States is currently Saudi Arabia's leading arms supplier. Total value of arms agreements with Saudi Arabia from 1950 through March 31, 1997, was $93.8 billion, while arms agreements with Saudi Arabia from 1994 through 2001 amounted to $4.6 billion. The upsurge in Saudi arms purchases from the United States after 1990 was due in large measure to the Persian Gulf crisis and its aftermath. The largest recent sale was a $9 billion contract for 72 F-15S advanced fighter aircraft, signed in May 1993.

As Table 1 shows, approximately 21% of the value of U.S.-Saudi arms contracts from 1950 to 1997 were for lethal equipment (i.e., weapons, ammunition, and combat vehicles, aircraft, and ships); the largest portion (32%) went for support services (repair, rehabilitation, supply operations, and training). Another major component of the Saudi program has been construction of military bases and facilities, accounting for 19%, although most military infrastructure projects were completed by 1990. A downward trend has marked Saudi arms procurement since the mid-1990s as Saudi Arabia completed many of its post-Gulf War purchases and the country faced straitened finances.

Table 1. U.S. Arms Transfers to Saudi Arabia, 1950-1997

Category	**Orders Deliveries**		**Deliveries**	
	$ in Billions	**% of Total Orders**	**$ in Billions**	**% of Total Deliveries**
Weapons & Ammunition	19.893	21.2	9.092	15.6
Support Equipment	16.614	17.7	9.815	16.8
Spare Parts & Modifications	9.778	10.4	5.259	9.0
Supply, Repair, Training	29.615	31.6	17.804	30.6
Construction	17.924	19.1	16.197	27.8
TOTALS	**93.824**	—	**58.167**	—

Source: DOD. All figures are current through March 31, 1997, which is latest detailed data available. U.S. arms *ordered* by Saudi Arabia from 1998 through 2001 amounted to $600 *million* (considerably lower than in previous years), while U.S. arms *delivered* to Saudi Arabia during the same period amounted to $12.8 *billion*, reflecting earlier orders.

Saudi-Pakistani Contacts

According to press reports in 1999, U.S. officials were concerned over a visit by the Saudi Defense Minister to nuclear and missile facilities in Pakistan, but had received assurances from Saudi officials that Saudi Arabia was not seeking nuclear weapons or other weapons of mass destruction. Four years later, a *Washington Times* article of October 22, 2003, citing "a ranking Pakistani insider," reported that Saudi Arabia and Pakistan have concluded a secret agreement on nuclear cooperation, under which Pakistan would provide Saudi Arabia with nuclear technology in return for oil at reduced prices. On the same day, a U.S. State Department spokesman said the U.S. Administration had seen these reports but added that "[w]e have not seen, however, any information to substantiate what would seem to us to be rather bald assertions." The spokesman went on to note that Pakistan understands U.S. concerns over nuclear proliferation and that Saudi Arabia is a party to the Nuclear Non-Proliferation Treaty. The Saudi Ambassador to the United States denied that Saudi Arabia and Pakistan had concluded a nuclear agreement and repeated previous Saudi calls for a nuclear-free zone in the Middle East.

Contacts with China

In 1988, shortly before the end of the cold war, Saudi Arabia concluded a controversial purchase of approximately 30 intermediate range CSS-2 missiles from China, in its first and only major arms purchase from a communist (or formerly communist) state. A Reuters news wire report of February 15, 2004, quoted unnamed U.S. officials as voicing concern over continued alleged cooperation between China and Saudi Arabia on missiles; the report did not provide details. On the following day, the Saudi Press Agency said a responsible source at the Saudi Defense Ministry denied the report, which a defense spokesman described as "fabricated and baseless."

Trade Relationships

Saudi Arabia was the largest U.S. trading partner in the Middle East in 2003. For that year, Saudi exports to the United States were estimated at $17.7 billion (a marked increase from the previous year's 12.2 billion) and imports from the United States at $4.3 billion. Comparable figures for Israel, the second largest U.S. trading partner in the Middle East in 2003, were $12.8 billion in exports and $4.6 billion in imports . To a considerable extent, this high volume of trade is a result of U.S. oil imports from Saudi

Arabia and U.S. arms exports to that country. The Saudis buy significant amounts of U.S. commercial equipment, such as machinery, vehicles, and electronic items as well. Saudi Arabia has applied to join the 128-member World Trade Organization (WTO), and U.S. officials including President Bush have expressed support for its application.

Oil Production

With the world's largest proven oil reserves (estimated at 261.7 billion barrels in January 2001), Saudi Arabia produced an average of 8.8 million barrels per day (bpd) of crude oil as of May 2003. Approximately 14.5% of U.S. oil imports and 8.8% of total U.S. oil consumption came from Saudi Arabia during 2003. Formerly the largest foreign supplier of oil to the United States, Saudi Arabia has been exceeded in this role by Venezuela and/or Canada during recent years (see Table 2).

In recent years, Saudi Arabia has alternately supported cuts and increases in production as oil prices on the international market have fluctuated. Under a "gentlemen's agreement" reached in June 2000, members of the Organization of Petroleum Exporting Countries (OPEC) established a mechanism to adjust the supply of oil by 500,000 bpd if the 20-day average price of oil moved outside a \$22 to \$28 price band. Members disagree, however, as to whether this mechanism is automatic or requires separate action by OPEC to implement, and Saudi Arabia has spoken of a target price of \$25 rather than a price band. Prices have risen markedly in recent years as regional conflict and terrorism have fueled concern over oil supply, bringing prices to a peak of nearly \$50 per barrel on August 20, 2004, before receding to \$45 on September 2. Saudi Arabia, the only OPEC state with spare production capacity, raised production from 9.25 million barrels per day to 9.5 million barrels per day in August 2004 and may pump more in subsequent months.

Table 2. Oil Consumption and Imports (in millions of barrels per day)

Category	2000	2001	2002	2003
Total U.S. Consumption	19.701	19.649	19.761	20.044
Total U.S. Imports	11.459	11.871	11.530	12.254
Imports from Saudi Arabia	1.572	1.662	1.552	1.772
Imports from Venezuela	1.546	1.553	1.398	1.385
Imports from Canada	1.807	1.828	1.971	2.068

Source: DOE.

Foreign Investment

Saudi leaders, notably Crown Prince Abdullah, have shown increasing interest in attracting foreign investment, especially in their country's energy sector. Oil exploration and production remain generally off limits to foreign investment; however, with the world's fourth largest natural gas reserves (219.4 trillion cubic feet), Saudi Arabia is emphasizing foreign participation in the country's gas sector. In 2001, Saudi Arabia signed three preliminary agreements worth approximately $25 billion with eight international oil companies (including six U.S.-based companies) to develop three natural gas fields. Conclusion of final agreements met continuing delays as Saudi and company negotiators tried to resolve several issues, including taxes, rate of return on investments, and size of gas reserves being offered. Eventually, in mid-2003, negotiations collapsed, and Saudi officials began pursuing a different approach involving smaller, less ambitious projects of more limited scope. Meanwhile, at a July 22-24 conference in London attended by over 40 companies including some U.S. companies, Saudi officials offered three other contract areas for exploitation and issued tenders on September 15. International oil companies were given until January 26, 2004, to submit bids.

Only six of the 40 companies that had expressed interest, including one U.S. company (Chevron-Texaco), submitted bids. The three winners, announced on January 26-28, 2004, were China's SINOPEC, Russia's Lukoil Holdings, and a consortium consisting of Italy's ENI SpA (E) and Spain's Repsol YPF S.A. (REP.MC). The state-owned Saudi Arabian Oil Company (Saudi ARAMCO) will hold a 20% share in each operating company. The U.S. Chevron-Texaco, which had bid on all three of the contract areas, was unsuccessful in obtaining any of the contracts. Saudi Oil Minister Ali al-Naimi and other officials said the bids were assessed strictly in accordance with the terms offered by the various companies, although some unnamed observers suggested that the bidding process was structured in a way that favored non-U.S. competitors. Still other observers suggested that the awards reflect Saudi desire to cement economic ties with Russia, China, and other third countries, and one energy expert noted that "[t]he Saudis are clearly shifting around and looking to different parts of the world.[17]

Human Rights, Democracy, and other Issues

Of particular concern to Westerners are pervasive restrictions on women's activities and an injunction against the practice of other religions throughout the Kingdom. This injunction has been applied not only against non-Islamic faiths but also at times against the Shi'ite Muslim community in Saudi Arabia, estimated at 500,000 or more persons mainly in the Eastern Province. Since 1990, the Saudi government has moved quietly to ease some restrictions on Shi'ites. A petition presented by Saudi Shi'ite representatives was followed by an audience with Saudi Crown Prince Abdullah on April 30, 2003, indicating that both the government and the Shi'ite petitioners may be seeking to pursue a more cooperative approach. Also, according to the State Department, high-level Saudi officials have said that Saudi policy allows for private non-Muslim worship, for example, in private homes or secluded compounds; however, in its annual report on human rights practices, published on February 25, 2004, the State Department notes that Saudi officials do not always follow these guidelines in practice. In the State Department's most recent annual Report on International Religious Freedom (September 15, 2004), Saudi Arabia is designated for the first time as a "Country of Particular Concern."

Political reforms promulgated by King Fahd in the early 1990s and continued under Crown Prince Abdullah appear to represent a limited move toward democracy and protection of individual freedoms. The "main law" announced by the King on March 1, 1992, bans arbitrary arrest, harassment, or entry of individual homes without legal authority and specifies privacy in telephone calls and mail. On August 20, 1993, the King appointed a 60-member consultative ("Shura") council (increased to 90 in 1997 and to 120 in 2001), with limited powers to question cabinet members and propose laws. On January 25, 2005, the Saudi Minister of Defense and Aviation and 2nd Deputy Prime Minister Prince Sultan told Saudi newspapers that the council will be expanded to 150 members and given additional unspecified authority in the next three months. In a further step, the Saudi government announced in August and September 2004 that municipal elections will be held for the first time since 1963 in a three-stage process, scheduled to begin on February 10, 2005. According to current plans, citizens will elect half the members of the country's 178 municipal councils. Campaigning began on January 30, 2005. Women reportedly will be excluded from these elections but may have the opportunity to participate in future ones.

BACKGROUND TO U.S.-SAUDI RELATIONS

Political Development

Saudi Leadership

As the birthplace of the Islamic religion in 622 A.D. and as the home of Islam's two holiest shrines (the cities of Mecca and Medina), the Arabian Peninsula has always occupied a position of special prestige within the Middle East. With the establishment of Arab empires based in Damascus and Baghdad, the peninsula gradually lost its political importance and sank into disunity. In the 16th century, much of the Arabian Peninsula came under the nominal rule of the Ottoman Empire; however, tribal leaders effectively controlled most of the region. During this period, an alliance developed between an influential eastern tribe, the House of Saud, and the leaders of a puritanical and reformist Islamic group known as the Wahhabi movement. During the first quarter of the 20th century, a chieftain of the Saud family, Abd al-Aziz ibn Abd al-Rahman (later known as Ibn Saud) overcame numerous rivals with the support of his Wahhabi allies and succeeded in unifying most of the Arabian Peninsula under his rule. Four sons have succeeded him.

Royal Succession

King Fahd, the current ruler and a dynamic leader for many years, is approximately 80 years old and suffers from medical problems including diabetes and arthritis. In early 1996, King Fahd temporarily turned over affairs of state to his half brother, Crown Prince Abdullah, for a six-week period while the King recuperated from a stroke. Amid conflicting reports about the King's condition, Crown Prince Abdullah has increasingly carried out many governmental functions since 1996, together with other senior princes of the royal family. Another key figure is Defense Minister Prince Sultan, a full brother of King Fahd, who is generally considered next in line of succession after Prince Abdullah. (King Fahd, Prince Abdullah, and Prince Sultan also hold the positions of Prime Minister, First Deputy Prime Minister, and Second Deputy Prime Minister, respectively.)

Most commentators believe the royal family would back Crown Prince Abdullah in a smooth transfer of power if King Fahd should pass from the scene. Various sources describe Prince Abdullah as more traditional and less western in outlook than King Fahd and more oriented toward the Arab world. On balance, the Crown Prince seems likely to maintain Saudi

Arabia's long-standing strategic and economic ties with the United States. U.S. officials commented that President Bush and Crown Prince Abdullah established a very good personal rapport during the latter's visit in April 2002. Some speculate, however, that succession could become more intricate after Abdullah (who is only two years younger than Fahd but believed to be in better health) and fear that future intra-family rivalries could weaken the Saud dynasty over the long term. Possible future candidates include some 25 brothers and half-brothers of King Fahd and a number of sons and nephews.

U.S. Aid, Defense, and Security

As Saudi oil income expanded, U.S. economic aid ended in 1959. Small amounts of aid continued through 1975, limited to a small international military education and training (IMET) program after 1968. Total U.S. aid to Saudi Arabia from 1946 through its termination in 1975 amounted to $328.4 million, of which $295.8 million was military and $32.6 million was economic assistance. Approximately 20% of total aid was in the form of grants and 80% in loans, all of which have been repaid. A small IMET program of $22,000 per year to help defray some expenses of sending Saudi officers to U.S. military service schools was resumed in FY2002 and increased slightly to $25,000 in 2003; $25,000 was requested by the Administration for Saudi Arabia in FY2004. This amount permits Saudi Arabia to purchase additional U.S. training at a lower cost than that which is charged to countries not eligible for IMET. Saudi officials also cite their country's role as a donor of aid to less affluent countries; according to the Saudi Embassy in Washington, Saudi Arabia gave $1.7 *billion* in development assistance and $14.7 *million* in relief assistance to developing countries in 2002.

Congressional Interest in Saudi Arabia

U.S.-Saudi bilateral issues of interest to Congress in recent years include arms sales to Saudi Arabia, Arab-Israeli issues including the now somewhat attenuated Arab boycott of Israel, oil pricing, Saudi trade practices including disputes between U.S. companies and Saudi clients (largely resolved), and — increasingly — terrorism. In the 108th Congress, H.R. 3137, introduced on September 17, 2003, would ban direct aid to Saudi Arabia along with five other named countries. In this regard, as mentioned above, Saudi Arabia only receives $25,000 in annual U.S. aid under the IMET program. Companion bills entitled the Saudi Arabia Accountability Act, introduced in the House and Senate on November 21 (H.R. 3643) and November 18, 2003 (S. 1888), respectively, would require the President to prohibit export of arms or dual use items to Saudi Arabia and to restrict the travel of Saudi diplomats in the United States, unless the President certifies that Saudi Arabia is fully cooperating with the United States in investigating and preventing terrorist attacks and taking other specified counter-terrorist measures.

During the debate on H.R. 4818, the Foreign Operations Appropriations bill for FY2005, the House adopted Amendment 708 introduced by Representative Anthony D. Weiner, which stated that "None of the funds appropriated or otherwise made available pursuant to this act shall be obligated or expended to finance any assistance to Saudi Arabia." This amendment, which the House passed by a vote of 217 to 191 (Roll no. 389), was included as Section 582 of H.R. 4818. Subsequently H.R. 4818 passed the House by 365-41 (Roll no. 390) on July 15, 2004. The Senate version of H.R. 4818, passed by voice vote on September 23, did not contain this ban. The Foreign Operations Appropriation bill was included as Division D of H.R. 4818, the Consolidated Appropriations Act for FY2005. The conference report on H.R. 4818 (H.Rept. 108-792) contained the ban in the earlier House version on aid to Saudi Arabia but provided waiver authority if the President certifies to the congressional appropriations committees that Saudi Arabia is cooperating in the war against terrorism and that U.S. assistance will facilitate that effort (Section 575). The President signed the bill as P.L. 108-447 on December 8, 2004.

On July 22, 2004, Senators Charles Schumer and Susan Collins introduced S.Con.Res. 131, calling on the government of Saudi Arabia to cease supporting religious ideologies that promote human rights abuses and urging the United States government to promote religious freedom in Saudi Arabia.

Section 4081(b) of the House-passed version of S. 2845, the 9/11 Commission Implementation Act of 2004 (also known as the National Intelligence Reform Act of 2004), would require the President to submit to designated congressional committees a strategy for collaboration with the people and government of Saudi Arabia including a framework for security cooperation. This strategy would include special reference to efforts to combat terrorist financing. The House passed this version of the bill on October 16, 2004 by 169- 229, Roll No. 525. The Senate-passed version also contains a reporting requirement on collaboration with Saudi Arabia (Section 1017(b)(4)); this version was passed by the Senate on October 6, 2004 by

96-2, Record Vote No. 199. Senate Amendment No. 3889, which would have established within the legislative branch a commission on the U.S.-Saudi relationship, was ruled non-germane by the chair. The conference report (H.Rept. 108-796, December 7, 2004) requires the President to submit to designated congressional committees a strategy for collaboration with Saudi Arabia, as part of a larger report on U.S. government activities to implement the provisions of this act (Section 7119(b)(3)). President Bush signed the bill on December 17, 2004, as P.L. 108-458.

REFERENCES

[1] Matthew Levitt, "Saudi Financial Counterterrorism Measures (Part II): Smokescreen or Substance?" *Policywatch No. 687*, Washington Institute for Near East Policy, Dec. 10, 2002, p. 1. A U.S. Treasury Department official confirmed on May 6, 2004, that the FIU had been established.

[2] "Classified Section of Sept. 11 Report Faults Saudi Rulers," *New York Times*, July 26, 2003.

[3] "Saudis Slam Congressional Report Accusing Kingdom of Poor Cooperation in Terror War," *Associated Press* News Wire, July 26, 2003.

[4] "Bush Refuses to Declassify Saudi Section of Report," *New York Times*, July 30, 2003.

[5] "Bush Vows 'American Justice' For Bombers in Saudi Arabia," *Washington Post*, May 14, 2003.

[6] See Craig Whitlock and Renae Merle, "Saudi Victims Had Military Link," *Washington Post,* June 14, 2004.

[7] "Ql Qaeda Shifts Its Strategy In Saudi Arabia," *The Washington Post*, Dec. 19, 2004.

[8] "U.S.-Saudi Anti-Terror Operation Planned," *Washington Post*, Aug. 26, 2003; "A Campaign to Rattle a Long-Ruling Dynasty," *New York Times*, Nov. 10, 2003.

[9] "Turning Back the Tide," *Middle East Economic Digest*, July 23-29, 2004, p. 23.

[10] See for example "U.S. And Saudis Agree On Cooperation," *Washington Post*, Feb. 26, 2003.

[11] "Saudis Challenge U.S. Over Iraq Fighters," *Washington Post*, Aug. 26, 2003.

[12] "U.S. Officers in Iraq Find Few Signs of Infiltration by Foreign Fighters," *New York Times*, Nov. 19, 2003.

[13] Tom Everett-Heath, "Opposing Views of the Kingdom to Come," *Middle East Economic Digest*, Jan. 23-29, 2004, p. 1.

[14] "Flow of Saudi Cash to Hamas Is Under Scrutiny by U.S.," *New York Times*, Sept. 17, 2003.

[15] Don Van Natta, Jr., with Timothy L. O'Brien, "Flow of Saudis' Cash to Hamas Is Scrutinized," *The New York Times*, Sept. 17, 2003. The report cites American law enforcement officials, American diplomats in the Middle East, and Israeli officials.

[16] Don Van Natta, Jr. with Timothy L. O'Brien, "Flow of Saudis' Cash to Hamas Is Scrutinized," *The New York Times*, Sept. 17, 2003.

[17] Karen Matusic, "Saudis Extend Geopolitical Base with Gas Deals," *The Oil Daily*, Feb. 2, 2004; Simeon Kerr, "Saudi Arabia Strikes Hard Bargain On Gas Rights," *Dow Jones, Emerging Markets Report*, Jan. 29, 2004.

In: Saudi Arabia: Terrorism, US Relations and Oil ISBN 1-59454-729-7
Editor: Nino P. Tollitz, pp. 27-61 © 2005 Nova Science Publishers, Inc.

Chapter 2

SAUDI ARABIA: TERRORIST FINANCING ISSUES*

Alfred B. Prados and Christopher M. Blanchard

SUMMARY

The September 11, 2001 attacks fueled criticisms within the United States of alleged Saudi involvement in terrorism or of Saudi laxity in acting against terrorist groups. Of particular concern have been reports that funds may be flowing from Saudi Arabia and other Middle East countries to terrorist groups, largely under the guise of charitable contributions. Critics of Saudi policies have cited a number of reports that the Saudi government has permitted or encouraged fund raising in Saudi Arabia by charitable Islamic groups and foundations linked to Osama bin Laden's Al Qaeda organization or like-minded entities.

The final report released by the bipartisan National Commission on Terrorist Attacks Upon the United States (the 9/11 Commission) indicates that the Commission "found no evidence that the Saudi government as an institution or senior Saudi officials individually funded [Al Qaeda]." The report also states, however, that Saudi Arabia "was a place where Al Qaeda raised money directly from individuals and through charities," and indicates that "charities with significant Saudi government sponsorship" may have diverted funding to Al Qaeda.

---

* Excerpted from CRS Report RL32499, dated December 8, 2004.

In numerous official statements and position papers, Saudi leaders have said they are committed to cooperating with the United States in fighting terrorist financing, pointing out that Saudi Arabia itself is a victim of terrorism and shares the U.S. interest in combating it. Saudi leaders acknowledge providing financial support for Islamic and Palestinian causes, but maintain that no Saudi support goes to any terrorist organizations, such as the Hamas organization. The U.S. State Department in its most recent annual report on international terrorism states that Hamas receives some funds from individuals in the Persian Gulf states but does not specifically mention Saudi Arabia.

Since the September 11 attacks, Saudi Arabia has issued numerous decrees and created new institutions designed to tighten controls over the flow of funds in or through the kingdom,. with particular emphasis on increasing the effectiveness of governmental supervision over charitable donations and collections. A rash of terrorist attacks on residential and office compounds in Saudi Arabia in 2003-2004 appears to have given added impetus to the Saudi leadership in expanding counterterrorist efforts. Since mid-2003, the Saudi government has: set up a joint task force with the United States to investigate terrorist financing in Saudi Arabia; shuttered charitable organizations suspected of terrorist ties; passed anti-money laundering legislation; banned cash collections at mosques; centralized control over charitable collections; closed unlicenced money exchanges; and scrutinized clerics involved in charitable collections. H.R. 10, introduced on September 24, 2004, would require the President to submit to designated congressional committees a strategy for U.S.- Saudi collaboration, with special reference to combating terrorist financing.

For more information about Saudi Arabia, see CRS Issue Brief IB93113, *Saudi Arabia: Current Issues and U.S. Relations*, by Alfred Prados. Audrey Kurth Cronin and Febe Armanios contributed to this report. This report will be updated periodically.

OVERVIEW

The attacks of September 11, 2001 fueled criticism within the United States of alleged Saudi involvement in terrorism or of Saudi laxity in acting against terrorist groups. One area of particular concern is the suspicion that public or private funds may be flowing from Saudi Arabia and other Middle East countries to finance international terrorist activity.

Reliable figures on the amount of money originating in or passing through Saudi Arabia and ending up in terrorist hands generally are difficult to obtain, for several reasons. First, the relatively small amounts of money required for terrorist acts can easily pass unnoticed. Second, the structure of the Saudi financial system makes financial transfers difficult to trace. Personal income records are not kept for tax purposes in Saudi Arabia and many citizens prefer cash transactions. Third, Muslim charitable contributions (*zakat*) are a religious obligation, constituting one of the five "pillars of Islam." Contributions are often given anonymously, and donated funds may be diverted from otherwise legitimate charities. Moreover, Saudi funding of international Islamic charities is reportedly derived from both public and private sources which in some cases appear to overlap, further complicating efforts to estimate the amounts involved and to identify the sources and end recipients of these donations.

This report reviews allegations of Saudi involvement in terrorist financing together with Saudi rebuttals, discusses the question of Saudi support for religious charities and schools (*madrasas*) abroad, discusses recent steps taken by Saudi Arabia to counter terrorist financing (many in conjunction with the United States), and suggests some implications of recent Saudi actions for the war on terrorism.

SAUDI ARABIA: ALLEGATIONS OF TERRORIST FINANCING

In the aftermath of the September 11 attacks, numerous allegations have been leveled against the Saudi Arabian government and prominent Saudi citizens regarding financial support for international terrorist groups. Although many of the allegations fault the Saudi government for failing to act decisively to close down channels of financial support, some critics go so far as to accuse Saudi government officials of responsibility for the September 11 attacks through design or negligence and for the continuing threat posed by the perpetrators or by like-minded terrorist groups. Some of the allegations also deal with possible ties between Saudi officials, private citizens, and the exiled Saudi terrorist Osama bin Laden and his Al Qaeda organization. However, Saudi authorities revoked bin Laden's citizenship in 1994, and senior Saudi officials have stated publicly that the Saudi regime is as much a target of Al Qaeda as is the United States.[1] There also have been suggestions that members of the large and sometimes fractious Saud family, numbering 5,000 or more, have acted independently of the country's senior leadership in providing support to Islamic fundamentalist groups prone to violence. Since September 11, some U.S. government officials have described Saudi Arabia as "a strong ally to the United States in [the fight against terrorist financing];"[2] however, others have qualified this viewpoint, and at least one has described Saudi Arabia as "the epicenter" of financing for Al Qaeda and other terrorist organizations.[3] The following are summaries of the more publicized post-September 11 reports of alleged Saudi involvement in terrorist financing.

General Allegations

The 9/11 Commission[4]

The final report released by the bipartisan National Commission on Terrorist Attacks Upon the United States indicates that the Commission "found no evidence that the Saudi government as an institution or senior Saudi officials individually funded [Al Qaeda]." The report also states, however, that Saudi Arabia "was a place where Al Qaeda raised money directly from individuals and through charities," and indicates that "charities with significant Saudi government sponsorship," such as the Al Haramain Islamic Foundation, may have diverted funding to Al Qaeda (For more on this subject see 'Action Against Questionable Charity' below).

Specifically, the report describes bin Laden's use of "the Golden Chain," an informal financial network of prominent Saudi and Gulf individuals originally established to support the anti-Soviet Afghan resistance movement in the 1980s. U.S. officials state that this network collected funds and funneled it to Arab fighters in Afghanistan, and later to Al Qaeda, using charities and other non-governmental organizations.[5] According to the Commission's report, Saudi individuals and other financiers associated with the Golden Chain enabled bin Laden and Al Qaeda to replace lost financial assets and establish a base in Afghanistan following their abrupt departure from Sudan in 1996. These activities were facilitated in part, the report argues, by the "extreme religious views" that exist within Saudi Arabia and the fact that "until recently" Saudi charities were "subject to very limited oversight."

Although the report highlights a series of unsuccessful U.S. government efforts to gain access to a senior Al Qaeda financial operative who had been detained by Saudi Arabia in 1997, the report credits the Saudi government with assisting U.S. officials in interviewing members of the bin Laden family in 1999 and 2000. The report argues that these meetings were integral to U.S. efforts to understand the role of bin Laden's personal wealth in the financing of Al Qaeda. As a result of this assistance, U.S. officials learned that Saudi government actions in the early 1990s had in effect divested bin

Laden of his share of the bin Laden family fortune, leading Al Qaeda to rely thereafter on a "a core group of financial facilitators" based in the Persian Gulf "and particularly in Saudi Arabia" for funding. The report also credits Saudi government actions following the May 2003 bombings in Riyadh which "apparently reduced the funds available to Al Qaeda — perhaps drastically."

Council on Foreign Relations Studies [6]

An independent task force sponsored by the Council on Foreign Relations has issued two reports that address terrorist financing and Saudi Arabia's alleged financial support for terrorism. The task force's October 2002 report strongly criticized what it asserted was Saudi financial support for international terrorist groups. For example, the report stated both in its summary and in the main body: "For years, individuals and charities based in Saudi Arabia have been the most important source of funds for Al Qaeda. And for years, Saudi officials have turned a blind eye to this problem." The authors argued that a connection between Saudi donors and Al Qaeda is logical for several reasons:

> Saudi Arabia possesses the greatest concentration of wealth in the region; Saudi nationals and charities were previously the most important sources of funds for the *mujahideen* [fighters against the Soviet occupation in Afghanistan]; Saudi nationals have always constituted a disproportionate percentage of Al Qaeda's own membership; and Al Qaeda's political message has long focused on issues of particular interest to Saudi nationals, *especially those who are disenchanted with their own government* [emphasis added].

The report grouped Saudi Arabia with Pakistan, Egypt, and other Gulf states and regional financial centers as "source and transit countries," implying that donations to terrorist causes not only originate in these countries but also that donations from other countries pass through them en route to terrorist organizations. The Saudi government responded by charging that the report's allegations were based on "false and inconclusive information."[7] A Treasury Department spokesperson reportedly described the report as "seriously flawed" because in the Department's opinion it did not describe new initiatives adequately.[8]

The task force's second report, released in June 2004, updates the first report and chronicles the steps the Saudi Arabian government has taken to strengthen its financial, legal, and regulatory systems and to combat terrorist financing since late 2002. According to the report, the Saudi government has "announced the enactment or promulgation of a profusion of new laws and regulations and the creation of new institutional arrangements that are intended to tighten controls over the principal modalities of terrorist financing." The report welcomes these steps and concludes that "on a comparative basis Saudi Arabia has recently taken more decisive legal and regulatory action to combat terrorist financing than many other Muslim states."

The second report also identifies several areas in which its authors argue Saudi authorities can do more to combat terrorist financing. It argues that additional action is essential "because of the fundamental centrality that persons and organizations based in Saudi Arabia have had in financing militant Islamist groups on a global basis." The report cites a lack of publicly available enforcement information as the basis for its questions about the Saudi government's commitment to implement its new terrorist financing laws and regulations. For example, the report specifically criticizes the Saudi government's "failure to punish, in a demonstrable manner, specific and identified leaders of charities found to be funneling money to militant Islamist organizations." Saudi authorities claim to have prosecuted five individuals "for terror financing" and frozen the assets "of number of other individuals."[9]

In response to the second report's release, Adel Al Jubeir, foreign policy advisor to Saudi Crown Prince Abdullah and chief Saudi spokesman, charged that the task force's conclusions were "politically motivated, ill-informed, and factually incorrect."[10] A Treasury Department spokesperson reportedly agreed with the report's assertion that Saudi Arabia should take further steps to combat terrorist financing.[11]

The September 11 Lawsuit

In mid-August 2002, the families of more than 600 victims of the September 11 attacks filed a suit for approximately $1 trillion against three members of the Saudi royal family, a number of financial institutions and individuals, and the government of Sudan. The suit, which the Saudi media have described as an attempt to extort Saudi deposits in the United States, alleges that the defendants enabled the September 11 attacks to occur by making financial resources available to the perpetrators. On November 14, 2003, however, Judge James Robertson of the U.S. District Court of the District of Columbia ruled that two of the leading defendants, Defense Minister Prince Sultan and former Director of Intelligence Prince Turki, were entitled to foreign sovereign immunity for their official acts. [Burnett v. Al Baraka Inv. and Development Corp., 292 F.Supp.2d 9 (D.D.C. 2003)][12]

Linde V. Arab Bank

On July 2, 2004, six families of victims killed or injured in terrorist attacks in Israel filed a civil lawsuit against Arab Bank PLC of Jordan in the U.S. District Court of New York seeking $875 million in damages. Although the Saudi government is not a party to the lawsuit, the complaint alleges that the Saudi Committee for the Support of the Al Quds Intifada used accounts established at Arab Bank as conduits for funds to charities and individuals in the West Bank and Gaza Strip associated with the Islamic Resistance Movement (Hamas), Islamic Jihad, and other terrorist entities (for a full description of the Committee, see 'Funding for Palestinian Organizations' below). The complaint also alleges that the Saudi Committee used accounts in Arab Bank and Arab Bank branches in the Palestinian territories to provide "insurance benefits" to the families of suicide bombers and others killed or detained in confrontations with Israeli security forces. In 2003, the Central Bank of Jordan ordered all Jordanian banks "to freeze any dealing" with six Hamas figures and associated charities. In response to the lawsuit Shukri Bishara, a senior executive at Arab Bank, reportedly said that "the accusations are completely false and baseless and Arab Bank never participated in any activity knowingly that could lead to violence."[13] A Saudi embassy official also reaffirmed that the Saudi government has not provided financial support for terrorism.[14] [Linde et al. v. Arab Bank, 04 CV 02799 (E.D.N.Y. filed July 2, 2004)]

Scrutiny of Riggs Bank

The Riggs National Bank, which for two decades had been the chief banker for the Royal Embassy of the Kingdom of Saudi Arabia in Washington, has come under scrutiny from U.S. Treasury and law enforcement officials and has been fined $25 million "for numerous violations of the Bank Secrecy Act."[15] In 2003, Riggs began filing "suspicious activity reports" (SARs) of unusual or unexplained transactions involving the Saudi Embassy's bank accounts. Saudi accounts at Riggs were closed in early March 2004 at the bank's initiative, according to some sources, but Saudi sources say it was a Saudi decision to end the relationship.[16] On May 13, 2004, federal regulators fined Riggs $25 million for failing to actively monitor financial transfers through Saudi Arabian and other bank accounts.[17] The FBI is currently investigating the activities of the former chief federal bank examiner for Riggs from 1998 to

2002, the period in which alleged money laundering violations occurred. In 2002, the official left the Treasury Department to become a vice president at Riggs.[18]

Joint Congressional Report

The declassification and release in mid-2003 of the report of the Joint Inquiry into Intelligence Community Activities before and after the Terrorist Attacks of September 11, 2001 also brought attention to the alleged role of Saudi Arabia in supporting terrorism. In the 900-page report, a crucial chapter on foreign support for the hijackers is virtually all blanked out — 28 pages in all — because the Administration refused on national security grounds to release the chapter in a public forum. There has been speculation about the degree to which the deleted pages might reveal Saudi complicity in the September 11 attacks. According to the press, persons who claim to have read the still-classified section of the report say it covers Saudi links with individuals involved in the attacks.[19] The Saudi foreign minister appealed to President Bush to publish the censored pages so as to enable Saudi Arabia to rebut these suspicions, but the President refused on the grounds that an ongoing investigation of the September 11 attacks might be compromised.[20]

The Brisard "U.N." Report

A controversial private report issued by French investigator Jean-Charles Brisard in December 2002 made several detailed allegations about the involvement of prominent Saudi nationals in the financing of international terrorist organizations. The United Nations Security Council did not solicit or endorse Brisard's report, although it has been mischaracterized as a U.N. report in the public record.[21] Although most public allegations of Saudi support for terrorist activities have not quantified the amounts of money involved, Brisard's report asserted that Al Qaeda received between $300 million and $500 million during the decade prior to 2002, by "abusing this pillar of Islam [charitable donations] and taking advantage of the Saudi regulatory vacuum."[22] Brisard described Saudi donors as "wealthy businessmen and bankers." The report has been the subject of a lawsuit in the United Kingdom regarding defamation of character, and a British High Court ruled in July 2004 that allegations contained within the report were "untrue."[23] [Mahfouz v. Brisard &

Others, High Court of Justice, Queen's Bench Division, United Kingdom - [2004] EWHC 1735 (QB)]

Allegations of Support for Iraqi Insurgents

In October 2004, an unidentified Defense Department official told the press that private Saudi individuals and charities were channeling funds to insurgent groups in Iraq.[24] Saudi officials vigorously denied the claims and appealed for U.S. officials to provide concrete information in support of the charges so that Saudi authorities could investigate and prosecute any individuals or entities that may have been involved.[25] In December 2004, press reports cited intelligence gathered following U.S. military operations, including the November 2004 assault on Fallujah, which indicated that "a handful of senior Iraqi Baathists operating in Syria are collecting money from private sources in Saudi Arabia and Europe" and are channeling it to insurgent groups.[26]

Saudi Arabia has contributed $1 billion in loans and credits to reconstruction efforts in Iraq, and U.S. officials have engaged Saudi authorities regarding forgiveness of Iraqi debt.[27] The Saudi government administers official aid programs in Iraq under the supervision of the Saudi Committee for the Relief of the Iraqi People and the Saudi Red Crescent Society. Prince Nayef bin Abd Al Aziz, the Saudi Minister of Interior, directs the Committee's operations. The Committee held a fundraising telethon in April 2003, which reportedly raised $11.5 million for relief efforts in Iraq.[28] In May 2003, the Saudi government established an official government bank account, Unified Account Number 111, to consolidate public and private donations in support of the Iraqi people.[29] Saudi press reports describing the operations of the Committee and the Red Crescent Society indicate that numerous shipments of food aid and medical supplies have been delivered across Iraq since April 2003, including shipments to the former-insurgent stronghold of Fallujah during the summer of 2004.[30]

FUNDING FOR PALESTINIAN ORGANIZATIONS

Support for Palestinian causes and the provision of humanitarian aid to Palestinians has long been an important component of Saudi foreign policy, and many Saudis identify strongly with the Palestinian people and view support for Palestinian causes as a religious, cultural, or, in some cases, political obligation. Repeated allegations made by Israeli and Western sources have contended that Saudi support for Palestinian institutions and individuals has directly or indirectly supported Palestinian terrorist groups, although public reporting has not conclusively linked official Saudi government support to Palestinian terrorist organizations.

Saudi Arabia, like other Arab states, recognizes the Palestine Liberation Organization (PLO) as the legitimate representative of the Palestinian people and provides some financial support to Palestinian institutions sponsored by the PLO. A 2002 Saudi government report stated that overall government and private aid to the Palestinians had reached $2.61 billion.[31] Currently, Saudi officials say that government support to Palestinian causes, approximately $80 million to $100 million per year, goes solely to the Palestinian Authority, which was established pursuant to the Israeli-Palestinian agreement of September 13, 1993, known as the first Oslo Accord.[32]

Officially Sanctioned Relief Efforts

Saudi Committees

Within Saudi Arabia, two official charitable committees have solicited and delivered aid to Palestinian institutions, individuals, and causes since the onset of the second Palestinian intifada in October 2000: the Saudi Popular Committee for Assisting the Palestinian Mujahideen and the Saudi Committee for the Support of the Al Quds Intifada. According to Saudi press reports, Prince Salman bin Abd Al Aziz, the governor of Riyadh Province, established the Popular Committee and has directed its operations.[33] The Popular Committee's periodic public reports indicate that it provided approximately $8.8 million to the PLO from October 2000 to April 2003.[34]

The Saudi Committee for the Support of the Al Quds Intifada has served as the main conduit for Saudi financial and material aid to the Palestinian territories since its establishment under Royal Decree 8636 on October 16, 2000. Prince Nayef has directed the Committee's operations since its establishment. A December 31, 2003 report issued by the Saudi Embassy in Washington stated that the costs of the Al Quds Intifada Committee's 31 relief programs amounted to 524,265,283 Saudi riyals (SR) [$139,804,075], in addition to the costs of then-current projects, which amounted to a further 203,699,433 SR [$54,302,249]. The report also stated that "the value of the services provided by the Committee to the Palestinian people" through

December 2003 was equal to an additional 727,964,716 SR [$194,123,924].[35]

The Al Quds Intifada Committee's periodic public reports describe millions of riyals in monetary aid that its programs have provided to the Palestinian people in the form of financial transfers to the Palestinian Authority (PA), donations to charitable organizations in the West Bank and Gaza Strip, and by means of direct assistance to over 35,000 needy Palestinian individuals. In addition to financial assistance, the Al Quds Intifada Committee also has provided food, blankets, medicine, ambulances, and other aid in kind through programs aimed at supporting health care, education, and the provision of basic social services in the Palestinian territories that have been disrupted during the current uprising. The Committee has also constructed hundreds of homes for Palestinians that have been left homeless due to the ongoing violence. Palestinian officials have reportedly praised the Committee's support for the Palestinian people.

In June 2004, the Saudi government announced that the future activities of all Saudi charitable committees and organizations that send aid abroad (including "the Palestinian committees") will be monitored and directed by the Saudi Nongovernmental National Commission for Relief and Charity Work Abroad.[36] As of December 2004 it was unclear if this National Commission was operational or if either of the two Committees described above continued to operate.

Relief Coordination

Since 2000, both Committees have issued public solicitations encouraging Saudi citizens to make donations to support the welfare of the Palestinian people. In one often cited instance, the Al Quds Intifada Committee organized a telethon sponsored by King Fahd in April 2002 that raised over $110 million for families of Palestinians killed or injured in the uprising.[37] Saudi officials told their U.S. counterparts that the proceeds of this telethon were funneled through non-governmental organizations to provide humanitarian support to needy Palestinian families.

Unified Accounts

Saudi press reports indicate that the Al Quds Intifada Committee consolidated the proceeds of its fund-raising efforts along with public and private donations in support of Palestinian causes in national, government-sponsored unified accounts established in a number of Saudi banks.[38] "Unified Account Number 98" and "Unified Account Number 90" are referred to in Saudi press reports describing the Al Quds Intifada

Committee's activities, although it is unclear if both accounts existed simultaneously, if they are synonymous, if one superceded the other, or if either still exists.[39] According to Saudi press reports, Saudi citizens were encouraged to make donations to these accounts at their local banks.

Delivery Mechanisms

Once collected, the proceeds of the Al Quds Intifada Committee's fundraising efforts were delivered to Palestinian beneficiaries in a variety of ways. Material aid collected by the Committee, such as food, clothing, blankets, and vehicles, was delivered in cooperation with third parties such as the Jordanian Red Crescent Society and the United Nations Relief and Works Agency (UNRWA). The Committee also developed a special process to transfer financial aid to Palestinian individuals in cooperation with Jordan's Arab Bank PLC.[40]

In an April 2002 interview Dr. Sa'id Al Urabi Al Harithi, the "chairman of the Executive Committee in Support of the Al Aqsa Intifada" and an advisor to Prince Nayef, provided details about the process used to identify beneficiaries and deliver financial aid to them:[41]

- Using information provided by "welfare societies" and "official sources" in the Palestinian territories, the Al Quds Intifada Committee prepared lists of potential beneficiaries drawn from the ranks of "the wounded, the families of martyrs, the families of prisoners and disabled persons," and families that had been affected by the uprising. The word "martyr" is used to refer to those killed as a result of violence (usually Israeli actions). [42]
- Then, a "central committee" conducted a "general study" of the lists and verified "the names of the beneficiaries, their telephone numbers, social conditions, addresses, and the date and type of injury." According to Al Harithi, the Committee would "coordinate with the Palestinian Authority and the [Palestinian] ambassador on the information in [the lists]" to confirm that it was "precise, correct, and clear."
- Once the information had been confirmed and approved, the Committee, in coordination with "the National Arab Bank,"[43] opened a bank account in the beneficiary's name, into which a standard amount of riyals or dollars were transferred based on the individual's circumstances and the criteria set by the Committee for each of its different programs. For example, "the committee

allocated 20,000 riyals [$5,333] for the family of each martyr," and "the sum of 10,000 Saudi riyals [$2,666] for every prisoner."

- After the deposit had been made, a team in the Palestinian territories followed up "on the process of delivering aid directly to the beneficiaries" and filed "regular reports to the Saudi Committee in Support of the Al Quds Intifada."

Allegations of Support for Palestinian Terrorists

Allegations of Support for the Families of Suicide Attackers

In May 2002, Israeli officials released a report that alleged that the Saudi Committee for the Support of the Al Quds Intifada had "transferred large sums of money to families of Palestinians who died in violent events, including notorious terrorists."[44] The report argued that this alleged financial support encouraged Palestinian terrorism by easing the potential burden on the families of attackers. Saudi officials called the allegations "baseless and false," and stated "unequivocally that Saudi Arabia does not provide financial support to suicide bombers or their families."[45] Saudi officials also questioned the claim that Palestinian suicide attacks were financially rather than politically motivated. Statements made by Committee figures in response to specific claims that the Al Quds Intifada Committee provided support to the families of suicide bombers or otherwise supported terrorism have been less consistent.[46] Saudi government spokesman Adel Al Jubeir has discussed the possibility that money from the Committee may have gone to the families of suicide bombers, but has categorically ruled out the existence of any quid pro quo arrangement or reward system similar to that sponsored by former Iraqi President Saddam Hussein.[47]

Saudi Committee Website

The Al Quds Intifada Committee website, currently maintained in the name of "the Saudi Committee for Relief of the Palestinian People,"[48] contains over 40,000 transaction records that feature the names individuals who received humanitarian aid and financial support from the Committee. The website states that "many accounts were opened for the harmed persons at the Arab [B]ank branches in the Palestinian territory, and fixed aids [i.e. donations] were transferred to the harmed people in their respective accounts."[49] One of the Committee's programs supported Palestinian families whose primary breadwinners were killed by Israeli forces or under other violent circumstances during the uprising. Records for this program contain the names of deceased individuals, transaction numbers, the date of their deaths, their home towns, and the circumstances in which they were killed. The Committee's description of the program indicates that payments of 20,000 SR [$5,300] were transmitted to the family members of these individuals in their names following their death.

Of the 1,300 names contained in the records for this specific Committee program, over 60 match or closely resemble the names of known Palestinian militants who carried out attacks on Israeli military personnel and civilians from October 2000 to March 2002.[50] These individuals include suicide bombers and gunmen who were killed during actual and attempted attacks inside Israel and the Palestinian territories. The following is a sample of five names of beneficiaries featured on the Committee website which match those of individuals associated with attacks:[51]

- Said Hassan Hussein Hotari - identified as suicide bomber in June 1, 2001, attack on Dolphinarium nightclub in Tel Aviv. Hamas claimed responsibility.[52]
- Izzedin Shahil Ahmed Masri - identified as suicide bomber in August 9, 2001 attack on Sbarro pizza restaurant in Jerusalem. Hamas claimed responsibility.[53]
- Maher Muhiaddin Kamel Habeishi - identified as suicide bomber in December 2, 2001, attack on Haifa bus. Hamas claimed responsibility.[54]
- Wa'fa Ali Khalil Idris - female, identified as suicide bomber in January 27, 2002 street attack in Jerusalem. The Al Aqsa Martyrs Brigade claimed responsibility.[55]
- Mohammed Ahmed Abdel-Rahman Daraghmeh - identified as suicide bomber in March 2, 2002 attack on Orthodox Jewish neighborhood in Jerusalem. The Al Aqsa Martyrs Brigade claimed responsibility.[56]

Most of the Committee records assigned to individuals whose names match or closely resemble those of suicide attackers list "assassination" as the cause of death — however other records credit "martyrdom." A handful of these records indicate that the individuals they refer to were killed during a "martyrdom operation."[57]

Support To Hamas

Since the early 1990s, there have been unsubstantiated reports of Saudi public and private assistance to the fundamentalist Islamic Resistance Movement (Hamas), which the U.S. government has designated as a foreign terrorist organization.[58] The Saudi government has not officially described Hamas as a terrorist organization.

In its annual report on terrorism for the year 2001 (*Patterns of Global Terrorism, 2001*), the State Department noted that Hamas receives funding from "private benefactors in Saudi Arabia and other moderate Arab states." The 2002 and most recent 2003 editions of the report do not mention Saudi Arabia as a specific source of funding for Hamas. Other reports have estimated varying amounts of aid to Hamas from private donations. According to one report, individuals in Saudi Arabia have contributed approximately $5 million to Hamas each year, or approximately half of its annual operating budget.[59] Saudi spokesman Adel Al Jubeir rejoins that "no Saudi government money goes to Hamas, directly or indirectly."[60] Al Jubeir has been quoted as saying that he considers it "very likely" that "some Saudi individuals" have provided financial support to Hamas.[61]

Recently, reports citing unidentified U.S. and Israeli intelligence officials have indicated that Saudi funding for Hamas has been curtailed and replaced by other regional sponsors. In June 2004 testimony before the Senate Governmental Affairs Committee, former Treasury Department General Counsel David Aufhauser quoted "informed intelligence sources" as saying that "for whatever reason, the money going to Hamas from Saudi Arabia has substantially dried up."[62] Aufhauser indicated that Saudi financial support "has been supplemented by money from Iran and Syria flowing through even more dangerous rejectionist groups in the West Bank."[63] Similarly, a report in the June 23, 2004 edition of the Israeli daily *Maariv* quoted an unidentified Israeli military official as saying that "for the first time in years the Saudis have begun to reduce the flow of funds to Hamas and to the Gaza Strip."[64] This unidentified source attributes this change largely to U.S. pressure on Saudi Arabia to stem the flow of funding to Hamas and other terrorist organizations.

CHARITABLE GIVING AND *MADRASAS*

Charitable giving (zakat) is a religious obligation for Muslims, constituting one of the five "pillars of Islam." Many wealthy Saudis contribute approximately 2.5 percent of their annual income to charitable causes and relief organizations that fund religious education programs, orphanages, hospitals, and other development projects both within Saudi Arabia and around the world. One expert estimates Saudi charitable donations, in general, to be about $3 billion to $4 billion annually, of which 10-20% is disseminated abroad.[65] Saudi officials estimate that $100 million in charitable donations are directed abroad each year.[66] It is unclear how much money is channeled from Saudi Arabian donations towards charitable endeavors in the United States, but some estimates suggest that $100 million has been donated over the last decade.[67] To what extent these donations have stemmed from the Saudi government or from private individuals is unknown.

Charity Oversight

In response to criticism and allegations of involvement in terrorist financing directed at Saudi Arabian charitable organizations, the Saudi government has taken a series of steps to provide greater oversight to charitable giving in the Kingdom. In December 2002, the government announced the creation of the High Commission for Oversight of Charities to provide assistance to Saudi Arabian charities in reforming their operations and improving their transparency.[68] At that time, Saudi officials also indicated that all Saudi charities had been audited; however, the results of those audits have not been made publicly available.

In May 2003, the Saudi government introduced new banking regulations that prohibited private charities[69] and relief groups from transmitting funds overseas, until further regulations could be instituted to ensure that the money would not be channeled to terrorist organizations.[70] The new banking regulations do not place similar restrictions on the operations of "multilateral" charitable organizations based in Saudi Arabia, such as the Muslim World League (MWL), the International Islamic Relief Organization (IIRO), or the World Assembly of Muslim Youth (WAMY), which actively raise funds among the Saudi population.[71]

In February 2004, King Fahd issued a royal decree establishing the Saudi Nongovernmental Commission on Relief and Charity Work Abroad.[72] The Commission was publicly reintroduced in June 2004, and described as "the sole vehicle" through which all private Saudi donations marked for international distribution will flow in the future.[73] Once the Commission is operational, Saudi charities with overseas operations will either be dissolved or have their assets consolidated under the control of the new Commission. The extent to which these government steps will be effective remains to be seen.

Action Against Questionable Charity

The Saudi government recently dissolved the Al Haramain Islamic Foundation, a large charity with links to the royal family, after years of sustained criticism and a series of joint U.S.-Saudi actions relating to its alleged involvement in terrorist financing. U.S. investigators have linked a former Al Haramain employee to the 1998 U.S. embassy bombing in Tanzania,[74] and a June 2, 2004 Treasury Department statement called Al Haramain "one of the principal Islamic NGOs providing support for the Al Qaida [variant transcription of Arabic word] network and promoting militant Islamic doctrine worldwide."[75] Since March 2002, the United States and Saudi Arabia have jointly designated 11 overseas branches of Al Haramain as front organizations for terrorist activities. The two countries have asked the U.N. Al Qaeda and Taliban Sanctions Committee (which monitors sanctions pursuant to U.N. Security Council Resolution 1267) to add the branches to the Committee's consolidated list of terrorists tied to Al Qaeda and the Taliban.[76] U.S. authorities also designated Al Haramain's founder and director, Aqeel Abdulaziz Al Aqil, as a supporter of terrorism in June 2004.[77] In early January 2004, Al Aqil had stepped down from his position and remains in Saudi Arabia. Al Aqil's successor, Dabbas Al Dabbasi, resigned as director of the organization on July 14, 2004 because of "the freezing of the establishment's internal accounts and the inability to give charitable support."[78] Saudi officials indicate that Al Haramain's international operations will be absorbed by the new Nongovernmental Commission for Relief and Charity Work Abroad.

Saudi Support to Religious Schools (*Madrasas*)[79]

Madrasas are religious schools that have historically existed throughout the Muslim world. Their curricula varies regionally and culturally; however for the most part these schools provide a religiously-based education, focusing on the Quran and other Islamic texts, usually to students at the primary or secondary school age. *Madrasa*s offer a free education, room, and board to their students, and thus they appeal to impoverished families and individuals. On the whole, these religious schools are supported by private donations from Muslim believers through *zakat*.

On a global front, concern has been expressed over the spread of radical Islam through Saudi-funded schools, universities, and mosques, which exist in many countries including Bangladesh, Bosnia-Herzegovina, Indonesia, Pakistan, Uzbekistan, Spain, and even in the United States. Some view the teaching of Saudi Wahhabism, an Islamic movement that encourages a return to the pure and orthodox practice of the "fundamentals" of Islam, as threatening the existence of more moderate beliefs and practices in other parts of the Muslim world. In November 2003, a Saudi Embassy spokesman indicated in a press release that "we do not fund the so-called radical *madrassas* [variant transcription of Arabic word] that people accuse us of funding, because that goes against our policy."[80] To date, there are no published reports on the aggregate amount of funding which has been donated from inside Saudi Arabia to specifically support the building of *madrasa*s worldwide.

Saudi Counter-Terrorism Efforts

Post-September 11 Actions

Saudi officials maintain that they are working closely with the United States to combat terrorism, which they say is aimed as much at the Saudi regime as it is at the United States. A month after the September 11 attacks, the Saudi Government announced that it would implement U.N. Security Council Resolution1373, which called for freezing terrorist-related funds. The Saudi Arabian Monetary Agency has issued a number of "white papers" detailing actions taken by Saudi Arabia to combat terrorist financing. The most recent update, issued in April 2004, addresses various steps including adopted legislation, implemented regulations and resolutions, and cooperation with the United States.[81]

In the past, U.S. officials, while acknowledging Saudi efforts, have expressed frustration with Saudi reluctance to share information gleaned from Saudi investigations of terrorist incidents and to move against organizations and individuals suspected of involvement in terrorism. Since mid-2003, however, the Saudi government seems to have become increasingly convinced of the seriousness of the terrorist threat and its attendant financing. Many commentators attribute this increased Saudi concern to terrorist bombings of compounds in the Saudi capital of Riyadh in

May and November 2003, both linked by U.S. and Saudi officials to Al Qaeda.[82] Two more attacks in the spring of 2004 in which most victims were non- U.S. support the theory that the continuing attacks are aimed primarily at destabilizing Saudi Arabia. At a hearing before the Senate Select Committee on Intelligence on February 24, 2004, then-Director of Central Intelligence George Tenet stated that "since the May 12th bombings, the Saudi government has shown an important commitment to fighting Al Qaida in the kingdom, and Saudi officers have paid with their lives." The State Department's most recent issue of *Patterns of Global Terrorism* states that the May and November 2003 attacks "galvanized Riyadh into launching a sustained crackdown against Al Qa'ida's presence in the Kingdom and spurred an unprecedented level of cooperation with the United States."

Joint Task Force

Shortly after the May 2003 attacks in Riyadh, a joint U.S.- Saudi intelligence task force was set up to help identify the perpetrators. This set the stage for a more permanent bilateral group with a broader mission. During the first week in August 2003, a delegation of senior U.S. counter-terrorism officials from the National Security Council, the State Department, the Treasury Department, and the Federal Bureau of Investigation met with Saudi leaders reportedly to urge them to do more to cut off the funneling of money to terrorists through Saudi businesses and organizations.[83] U.S. officials traveled to Riyadh later that month to set up a joint task force to investigate terrorist financing in Saudi Arabia. Subsequently, agents from U.S. organizations including the Federal Bureau of Investigation and the Internal Revenue Service joined Saudi officials to set up a center that will focus on bank accounts, computer records, and other financial data.

The U.S.-Saudi task force apparently has two components: one focused mainly on intelligence and the other on financing. The intelligence side shares intelligence information related to threats to the United States, especially general criminal terrorism leads. The financing side shares information related to terrorist funding, including sharing financial leads, requests for bank records, information on accounts, and so on.[84] The overall task force is composed of a small groups of FBI agents, FBI employees, analysts, IRS agents and intelligence agents. Among other capabilities, it is apparently able to receive real time information on potential leads in the United States and then request Saudi assistance to pursue them in Saudi Arabia.[85]

According to press articles, this is the first time that U.S. law enforcement officials have been stationed in Saudi Arabia to pursue intelligence and financing issues.[86] Many observers see this joint effort as a test of how responsive Saudi Arabia may prove to be in dealing with the issue of terrorism and blocking the flow of money from its citizens to terrorist organizations.[87]

The FATF-GCC Assessment

In September 2003, members of the Financial Action Task Force (FATF)[88] and the Gulf Cooperation Council (GCC) visited Saudi Arabia to examine Saudi financial practices. The FATF 2004 annual report, released on July 2, 2004, states that Saudi Arabia's legal and regulatory system is "compliant or largely compliant with most of the FATF 40+8 Recommendations" on terrorist financing.[89] According to the report, Saudi Arabia has established a Permanent Committee on Combating the Financing of Terrorism to coordinate its policy response and the Saudi Anti-Financial Crime Unit (SAFCU) to serve as a clearinghouse for investigative information and international cooperation. The SAFCU was not operational at the time of the FATF review.

The FATF report concludes that the Saudi government's legal definition of terrorist financing "does not conform to the international standards as expressed in the UN International Convention on the Suppression of Terrorist Financing."[90] The FATF review dealt only with the adequacy of legal and regulatory provisions, and did not assess the degree to which these laws and regulations are being effectively implemented.

Other Recent Measures

Press reports, the State Department's *Patterns of Global Terrorism: 2003*, and statements by knowledgeable observers including a former Department of the Treasury official[91] have listed various recent measures taken by Saudi Arabia to put a stop to terrorist financing. Most of these measures have been designed to improve oversight of charitable collections, and include:

- A law adopted in August 2003 making money laundering and terrorist financing criminal offenses.[92]
- A ban on cash collections at mosques and on transfers abroad of charitable funds collected in Saudi Arabia, except with Foreign Ministry approval and subject to stringent reporting requirements.
- A requirement for charitable organizations to have single disbursement bank accounts and an approved official with signatory authority to facilitate tighter controls over such accounts.[93]

- Closure of unlicenced money exchange houses and closer supervision of informal money transfer houses used to send funds abroad, known as *hawalas*.
- New rules governing the insurance sector and capital markets, and establishment of a financial intelligence unit (the SAFCU) to collect and share information on suspicious financial transactions.[94]
- Vetting of religious clerics and supervision of money given to them by their congregations. The government suspended more than 1,000 clerics in 2003 and 900 clerics so far in 2004 supposedly "on the grounds of negligence."[95]
- Announcement in early December 2003 of a rewards program, ranging from $270,000 to $1.87 million, for information leading to the arrest of suspects or disruption of terrorist attacks.
- Announcement in late November 2004 that Saudi Arabia will participate in the newly established Middle East and North Africa Financial Action Task Force (MENAFATF). Membership in the regional body commits Saudi Arabia to implementing the internationally recognized anti-money laundering and counterterrorist financing standards designed by FATF.

There have also been specific developments in bilateral U.S.-Saudi cooperation in criminal investigations and procedures:

- In September 2003, the FBI and the Internal Revenue Service Criminal Investigative Division began a program of terrorism financing training for the Saudi government and has had sessions in Riyadh and Washington, D.C. A third session was scheduled for Riyadh in May 2004.[96]
- According to the U.S. Department of the Treasury, the Saudis have killed several known operatives, financial facilitators, and financiers for Al Qaeda, including Yousif Salih Fahad Al Ayeeri (a.k.a. "Swift Sword") and Khaled Ali Al Hajj, reportedly key financial facilitators for Al Qaeda in the Persian Gulf region.[97]

Continuing Uncertainties and Questions

Saudi Arabia has undertaken significant administrative reforms in its efforts to curtail terrorist financing. A FATF official involved in the recent assessment of Saudi practices was quoted as saying that the new regulations to control Saudi-based charities "probably go further than any country in the world."[98] As noted above, however, the effectiveness of many of these steps will depend on their implementation. Several uncertainties and unresolved questions remain.

The Khobar Towers Bombing

The bombing of a U.S. Air Force billeting facility in June 1996 has never been resolved. The reluctance of Saudi investigative officials to share evidence with their U.S. counterparts was a long-standing source of friction between the two governments; both issued their own separate indictments. Although the Saudi Hezbollah group and the Iranian government are suspected of involvement, recent reports indicate that Al Qaeda may also have played a role in the attack. Some observers believe that more and better U.S. access to Saudi files on this case might reveal money trails that could shed further light on the intricacies of terrorist funding and collaboration.

Dilatory Action

U.S. officials have complained in the past that Saudi officials have been slow to take action against organizations and entities implicated in terrorist financing. According to a monograph released by the 9/11 Commission, "the Saudi government turned a blind eye to the financing of al Qaeda" before September 11, 2001, and "the Saudis did not begin to crack down hard on Al Qaeda financing in the Kingdom until after the May 2003 Al Qaeda attacks in Riyadh."[99] The report recognized the "significantly higher levels of cooperation" that U.S. officials have received from their Saudi counterparts since May 2003. Securing timely cooperation remains an important component of U.S. efforts to curb the flow of funds to terrorist groups.

Accountability

U.S. officials have also said that the Saudi Government should give more emphasis to personal responsibility in reporting and act against those who tolerate or promote financing of terrorist activity. According to the 9/11 Commission Monograph on Terrorist Financing, "the Saudis have failed to impose criminal punishment on any high-profile donor."[100] The 2004 FATF review found that Saudi Arabia has initiated five criminal cases involving terrorist financing activities, with one successful conviction. Several prominent Saudi individuals suspected by U.S. authorities of involvement in terrorist financing have not been charged or prosecuted in the United States, Saudi Arabia, or other jurisdictions. These individuals are believed to be in Saudi Arabia.

Palestinian Funding

Saudi leaders have said that Saudi government contributions to Palestinian causes go exclusively to the Palestinian Authority[101] and not to Hamas. The State Department notes that Hamas receives funds from benefactors in various locations including the Persian Gulf, although it does not specifically mention Saudi Arabia. The possibility remains that private donors in Saudi Arabia may circumvent Saudi regulatory controls.

Currency Controls

Although Saudi Arabia's customer information requirements are more stringent for international transfers under recently adopted regulations, some loopholes may remain. The FATF report calls on Saudi officials to increase the amount of information required for foreign currency conversion transactions at money remittance centers in the Kingdom. Saudi law permits the undeclared transport of currency across its borders for all amounts under 100,000 Saudi *riyals* ($26,666).

International Efforts

The FATF review notes that Saudi Arabia has not ratified the International Convention for the Suppression of the Financing of Terrorism, adopted by the U.N. General Assembly in December 1999. The FATF report urges Saudi Arabia to do so "as soon as possible."

OUTLOOK

It seems clear that growing challenges to internal security and an increase in terrorist incidents in Saudi Arabia have impelled the Saudi leadership to devote heightened attention to countering sources of financing terrorism. This has been particularly true since mid-2003, when terrorists began mounting a series of attacks on residential and office compounds, apparently in an effort to target the Saudi government as well as the western presence in Saudi Arabia. Both U.S. officials and independent observers have welcomed the mechanisms that Saudi authorities have put in place with the aim of stemming the flow of funds destined for terrorist groups. They point out, however, that the effectiveness of these measures will be tested by the degree to which Saudi authorities succeed in implementing the various institutions and regulations that have been established in recent years.

Issues for Congress

Several Committees and Members have expressed interest in these issues and have held hearings and introduced legislation relating to U.S.-Saudi cooperation in the fight against terrorism and terrorist financing. Section 575 of the FY2005 Omnibus Appropriations Bill prohibits Saudi Arabia from receiving aid or any direct assistance from the United States. Saudi Arabia has been receiving $25,000 in annual U.S. aid under the International Military Education and Training (IMET) program. Section 575 provides the President with waiver authority for the aid prohibition provided that he certifies that "Saudi Arabia is cooperating with efforts to combat international terrorism and that the proposed assistance will help facilitate that effort." Companion bills entitled the Saudi Arabia Accountability Act, introduced in the House and Senate on November 21 (H.R. 3643) and November 18, 2003 (S. 1888), respectively, would require the President to prohibit export of arms or dual use items to Saudi Arabia and to restrict the travel of Saudi diplomats in the United States, unless the President certifies that Saudi Arabia is fully cooperating with the United States in investigating and preventing terrorist attacks and taking other specified counter-terrorist measures.

Following the release of the final report of the National Commission on Terrorist Attacks Upon the United States (the 9/11 Commission), Members of Congress introduced several bills to implement recommendations of the Commission with regard to the organization of the intelligence community and other foreign policy mechanisms. The 9/11 Recommendations Implementation Act (S. 2845, as CRS-25 amended by the House), for example, would require the President to submit to designated congressional committees a strategy for collaboration with the people and government of Saudi Arabia including a framework for security cooperation [Section 4081(b)]. This strategy would include special reference to efforts to combat terrorist financing.

References

[1] Foreign policy advisor to Saudi Crown Prince Abdullah, Mr. Adel Al Jubeir, said on *Meet the Press*, that "we are targeted by Al Qaeda .... Their objective is to change and topple the government in Saudi Arabia." Available at [http://saudiembassy.net/2003News/Statements/StateDetail.asp?cIndex=190]. In June 2004, Saudi Ambassador to the

U.S. Prince Bandar bin Sultan called in the Saudi press for "jihad" against Al Qaeda militants in Saudi Arabia. (Prince Bandar bin Sultan,"A Diplomat's Call for War," *Washington Post*, June 6, 2004.)

[2] U.S. Treasury Secretary John W. Snow, "Statement following Visit to Saudi Arabia," September 17, 2003.

[3] David D. Aufhauser, General Counsel, Department of Treasury, Testimony to Senate Judiciary Committee, June 26, 2003.

[4] National Commission on Terrorist Attacks Upon the United States, "The 9/11 Commission Report," July 22, 2004.

[5] FBI agents discovered a handwritten list of 20 alleged Al Qaeda financiers during a March 2002 raid on a Saudi-based charity in Sarajevo, Bosnia. Bin Laden apparently referred to the group as "the Golden Chain." Further details were submitted to the U.S. District Court for the Northern District of Illinois by federal prosecutors in January 2003. The 9/11 Commission Report cites the following document: "Government's Evidentiary Proffer Supporting the Admissibility of Co-Conspirator Statements, United States v. Enaam Arnaout, No. 02-CR-892 (N.D. Ill. filed Jan. 6, 2003)."

[6] *Terrorist Financing.* Report of an Independent Task Force Sponsored by the Council on Foreign Relations, October 2002. Available at [http://www.cfr.org/pdf/Terrorist_Financing _TF.pdf] and *Update on the Global Campaign Against Terrorist Financing.* Second Report of an Independent Task Force on Terrorist Financing Sponsored by the Council on Foreign Relations, June 15, 2004. See [http://www.cfr.org /pdf/Revised_Terrorist_Financing.pdf]. The Council on Foreign Relations does not institutionally endorse the studies produced by its independent task forces, and notes that task force members are solely responsible for the content of their respective reports.

[7] Royal Embassy of the Kingdom of Saudi Arabia, "Response to CFR Report," October 17, 2002. Available at [http://www.saudiembassy.net /2002News/Press/PressDetail.asp?cYear= 2002&cIndex=54].

[8] Douglas Farah, "Report: Terror Funds Flow Through Saudi Arabia," *Washington Post,* October 16, 2002.

[9] Adel Al Jubeir, Interview Transcript: *CNN Late Edition*, June 15, 2004.

[10] Adel Al Jubeir, "Saudi Arabia Blasts CFR Task Force Report," June 15, 2004. Available at [http://www.saudiembassy.net/2004News/Press /PressDetail.asp?cIndex=223].

[11] *Reuters*, "Saudis Doing Better Battling Terror Funding-Report," June 15, 2004.

[12] See also Carol D. Leonnig, "Judge Rejects Saudi Terrorist Link," *Washington Post*, Nov. 15, 2003.
[13] *Agence France Presse*, "Arab Bank Denies Allegations it Funded Terrorists," July 7, 2004.
[14] Personal communication with CRS analyst, July 13, 2004.
[15] See Office of the Comptroller of the Currency, "OCC Assesses $25 Million Penalty Against Riggs Bank N.A.," May 13, 2004. Available at [http://www.occ.treas.gov/scripts/ newsrelease.aspx?Doc=5AOFP8 K.xml].
[16] Timothy O'Brien, "A Washington Bank In a Global Mess," *New York Times*, Apr. 11, 2004; Kathleen Day and Terence O'Hara, "U.S. Ready To Fine Riggs Bank; Saudi Embassy Money Reports Scrutinized," *Washington Post*, Apr. 18, 2004 (corrected Apr. 27).
[17] Timothy L. O'Brien, "Regulators Fine Riggs $25 Million," *New York Times*, May 14, 2003.
[18] Kathleen Day, "Treasury D. Probing Former Riggs Examiner," *Washington Post*, June 17, 2004.
[19] James Risen and David Johnston, "Report on September 11 Suggests a Role by Saudi Spies," *New York Times*, Aug. 2, 2003.
[20] Mike Allen, "Bush Won't Release Classified September 11 Report," *Washington Post*, July 30, 2003, p. A1; David Johnson and Douglas Jehl, "Bush Refuses to Declassify Saudi Section of Report," *New York Times*, July 30, 2003.
[21] An April 29, 2004 correction appended to Joel Mowbray, "Saudis Behaving Badly," *National Review Online,* December 20, 2002, reads, "The Brisard Report was an unsolicited document submitted to the U.N. by its author." See also a private letter written by former U.N. Security Council President Alfonso Valdivieso that disassociates the U.N. from Brisard's report. Available at [http://www.binmahfouz.info /pdf/faqs_5_jsb_12march.pdf].
[22] Jean-Charles Brisard, *Terrorism Financing: Roots and Trends of Saudi Terrorism Financing.* JCB Consulting, Dec. 19, 2003. Available at [http://www.nationalreview.com/ document/ document-un122002. pdf].
[23] The full text of the Court's judgment is available at [http://www.binmahfouz.info/pdf/ faq_4_judgment.pdf].
[24] John Lumpkin, "Insurgents Infiltrating Iraq Have Cash," *Associated Press*, Oct. 22, 2004.

[25] Royal Embassy of Saudi Arabia, "Saudi Official Refutes Allegation that the Kingdom is Neglecting Issue of Funding to the Iraqi Insurgency," October 23, 2004.

[26] Thomas E. Ricks, "Rebels Aided By Allies in Syria, U.S. Says," *Washington Post*, December 8, 2004.

[27] Paul Richter, "Nations Slow to Deliver Iraq Aid," *Los Angeles Times*, July 12, 2004.

[28] Agence France Presse, "Saudi Telethon Raises $11.5 Million for Iraqis," Apr 28, 2003.

[29] See Royal Embassy of the Kingdom of Saudi Arabia, "Bank Account Ready to Receive Donations for Iraqi Aid," April 24, 2003. Available at [http://www.saudiembassy.net/2003News/ News/RelDetail.asp?cIndex=737]

[30] *Agence France Presse*, "Saudi Arabia Sends Aid to Residents of Iraqi Rebel Bastion," June 10, 2004; and *BBC Monitoring International Reports*, "Saudi Arabia Sends Relief to Iraq's Al-Fallujah," Saudi TV1 (Riyadh), August 8, 2004.

[31] Royal Embassy of the Kingdom of Saudi Arabia, "Kingdom's Aid to Palestinians (sic.) Nears Ten Billion Saudi Riyals," May 2, 2002. Available at [http://www.saudiembassy.net/ 2002 News/News/For Detail.asp?cIndex=1122]. It is unclear if this figure included contributions to the Saudi-sponsored, multilateral Al Aqsa and Al Quds Intifada (Jerusalem uprising) funds established by 22 Arab leaders in October 2000. See [http://www.saudiembassy.net/ 2000News/Statements/StateDetail.asp?cIndex=357].

[32] Don Van Natta, Jr, with Timothy L. O'Brien, "Flow of Saudi's Cash to Hamas Is Scrutinized," *New York Times*, Sept. 17, 2003, pp. A1, A10.

[33] In one press report, Prince Salman asserted that the Popular Committee had been established after the 1967 Arab-Israeli War "with the aim of assisting the Palestinians." It is unclear if the Popular Committee operated over a thirty year period or was reactivated in October 2000. See *Arab News* (Jedda) "Salman Rejects Terrorism Charges Against Charities," November 3, 2002.

[34] Data compiled from official and non-official Saudi press reports spanning the period from October 2000 to April 2003. See *Saudi Press Agency*, "PLO gets more than SR 1.8 million from the Popular Committee for Assisting the Palestinian Mujahideen," April 22, 2003, and Foreign Broadcast Information Service (FBIS) reports: "Press

Summary — Saudi Leadership Report,"Oct. 1-15, 2000 through June 16-30, 2004.

[35] Royal Embassy of the Kingdom of Saudi Arabia, "Humanitarian Relief Handed Over to Palestinian Officials in Kingdom," Dec. 31, 2003. See [http://www.saudiembassy.net/2003 News/News/ForDetail.asp?cIndex=1189].

[36] Adel Al Jubeir, "U.S. and Saudi Officials Hold a News Conference on a Major Development in the War on Terrorism," Federal Document Clearing House Transcript, June 2, 2004. The Commission's creation was originally announced in late February 2004.

[37] Royal Embassy of the Kingdom of Saudi Arabia, "Telethon for Palestinian Victims Highly Successful," April 14, 2002. See [http://www.saudiembassy.net/2002News/News/ForDetail.asp?cIndex=1132].

[38] *Saudi Press Agency*, "Prince Urges Donations for Jerusalem," October 10, 2000. The press report lists "the National Commercial Bank, the Saudi-American Bank, Al Rajhi Banking and Investment Corporation, the Saudi-British bank, the Saudi-Dutch Bank, the Arab National Bank, the Saudi-French Bank, and Al Riyadh Bank" as sponsors of Unified Account 98. FBIS Translations GMP20001006001246 and GMP20001021000192. See also Robert Lenzner and Nathan Vardi, "Terror Inc.," *Forbes*, October 18, 2004.

[39] Later, in an April 2002 interview, high-ranking Committee official Dr. Sa'id Al Urabi Al Harithi stated that "All relief agencies that collect funds in support of the Palestinians deposit the funds in the Committee's Account Number 90." FBIS Translations GMP20001006001246 and GMP20020419000070.

[40] The Committee's website and Committee officials (see below) state that the Al Quds Intifada Committee opened accounts in Arab Bank PLC, which operates 22 local branches throughout the West Bank and Gaza Strip. The Committee's website, [http://www.alquds-saudi.org], discusses the use of Arab Bank PLC branches as conduits to transfer funds to beneficiaries.

[41] "Saudi Al Aqsa Intifadah Committee Chief on Delivering Donations to Palestinians," *Jedda Ukaz*, April 2002. FBIS Translation - GMP20020419000070.

[42] "The Saudi government offers assistance to the families of those killed, and these, the victims of the violence, are often referred to as martyrs. Nevertheless, the Saudi government does not condone the act of killing oneself and killing others by means of suicide bombings, and

Saudi religious leaders have condemned the taking of innocent lives." See [http://www.saudiembassy.net/2002News /Statements/StateDetail .asp?cIndex=146]

[43] Jordan's Arab Bank PLC owns a 40% stake of Saudi Arabia's Arab National Bank. Although the Committee clearly states that funds were delivered via Arab Bank PLC branches it is unclear what role, if any, Saudi Arabia's Arab National Bank played in the process.

[44] Ben Barber, "Saudi Millions Finance Terror Against Israel," *Washington Times*, May 7, 2002. Israel also released allegedly seized documents that featured "the letterhead of the 'Saudi Arabian Committee for Support of the Al Quds Intifada.'"

[45] See Royal Embassy of the Kingdom of Saudi Arabia, "The Kingdom of Saudi Arabia Responds to False Israeli Charges," May 6, 2002. Available at [http://www.saudiembassy. net/2002News /Press/PressDetail.asp?cYear=2002&cIndex=36] and Royal Embassy of the Kingdom of Saudi Arabia, "Saudi Arabia Doesn't Reward Terrorism," May 11, 2002. Available at [http://saudi embassy.net/2002News/News/ TerDetail.asp?cIndex=45]

[46] Dr. Sa'id Al Urabi Al Harithi stated in April 2002 that the Al Quds Intifada Committee had "nothing to do with terrorism" and "nothing to do with politics," and added that Saudi Arabia "rejects and fights terrorism." [FBIS Translation GMP20020419000070] In May 2002, Mubarak Al Biker, the "executive manager" of the Al Quds Intifada Committee, stated "we support the families of Palestinian martyrs, without differentiating between whether the Palestinian was a bomber or was killed by Israeli troops." Ra'id Qusti, "Saudi Telethon Funds Go Direct to Palestinian Victims," *Arab News* (Jedda), May 27, 2002. Available at [http://www.arabnews.com/ ?page=1&section=0&article =15591&d=27&m=5&y=2002].

[47] See Royal Embassy of the Kingdom of Saudi Arabia, "Saudi Arabia Does Not Support Terrorism," Interview Published in *National Journal*, May 11, 2002. Available at [http:// [http://www.saudiembassy.net/2002News/Statements/TransDetail]s Said to Get Iraq Millions," *Boston Globe*, May 30, 2002.

[48] The website [http://www.alquds-saudi.org/] cites Royal Decree 8636 as the basis for the establishment of the "Relief Committee."

[49] See [http://www.alquds-saudi.org/programs/programsDetails.asp?id= 1].

[50] The Palestinian Human Rights Monitoring Group maintains comprehensive English and Arabic language lists of suicide attackers.

These lists contain names that also appear on the Saudi Committee website. Further information provided in these and other Committee records such as individuals' home towns and dates of death matches information provided in public press accounts of attacks. In some instances, date and name data reflect slight variations and inconsistencies. The PHRMG lists are available at [http://www.phrmg.org/PHRMG%20Documents/Suicide%20bombers/Tables/]

[51] The names listed are transcriptions of a sample of five names that appear on the referenced links to the Committee website. Transcriptions of Arabic names often differ in Western media reporting and there is considerable uniformity in Arabic names. Mohammed and Ahmed, for example, are very common Arabic names often found among Palestinians.

[52] See [http://www.alquds-saudi.org/programs/beneficiaryDetails.asp?id=96908] and Jamal Halaby, "Tel Aviv Suicide Bomber Was Muslim," *Associated Press*, June 3, 2001. Hamas record available at [http://www.palestine-info.info/arabic/hamas/shuhda/alhotri/alhotri.htm]

[53] See [http://www.alquds-saudi.org/programs/beneficiaryDetails.asp?id=96909] and Greg Myre, "Jerusalem Suicide Bomber Kills 15," *Associated Press*, August 10, 2001. Hamas record available at [http://www.palestine-info.info/arabic/hamas/shuhda/ ezaldenalmasre/almasre.htm]

[54] See [http://www.alquds-saudi.org/programs/beneficiaryDetails.asp?id=96972] and Karin Laub, "Islamic Militants Unleash Wave of Terror Against Israel," Associated Press, December 2, 2001. Hamas record available at [http://www.palestine-info.info/arabic/hamas /shuhda/maherhobashe/maher.htm]

[55] See [http://www.alquds-saudi.org/programs/beneficiaryDetails.asp?id=97020] and Hadeel Wahdan, "Family Describes Palestinian Bomber," *Assoc. Press*, January 30, 2002.

[56] See [http://www.alquds-saudi.org/programs/beneficiaryDetails.asp?id=96140] and Greg Myre, "Suicide Bomber Kills Nine in Jerusalem," *Associated Press*, March 2, 2002.

[57] The Arabic text in the "Details" field of these Committee records reads "amaliyah istishadiyah," a commonly used euphemism for a suicide attack that translates literally to "martyrdom operation." See, for example, [http://www.alquds-saudi.org/programs/ beneficiary Details.asp?id=96940] and [http://www.alquds-saudi.org/programs/beneficiaryDetails.asp?id=97139].

[58] Some individuals and groups in Europe, North America, and the Middle East have argued that the provision of financial support to Hamas is legitimate because of the social services the organization provides to Palestinians, which they argue are separate from the group's military wing, the Izzedine Al Qassam Brigades, and its role in conducting terrorist operations.

[59] Don Van Natta, Jr, with Timothy L. O'Brien, "Flow of Saudi's Cash to Hamas Is Scrutinized," *New York Times*, Sept. 17, 2003, pp. A1, A10.

[60] Ibid.

[61] *Agence France Presse*, "Saudi Official Condemns Terrorism, But Not Hamas," June 12, 2003.

[62] David Aufhauser, testimony before the U.S. Senate Committee on Governmental Affairs, June 15, 2004, transcript provided by Federal Document Clearing House.

[63] Ibid.

[64] Amir Buhbut, "Saudis Cutting Back Hamas Funding," *Maariv*, June 23, 2004. Available at [http://www.maarivintl.com/index.cfm?fuseaction=article&xCache={ts%20'2004-06-29 %20 21:43:06'}&articleID=8940].

[65] Jonathan M. Winer, Congressional testimony before the Committee on Senate Governmental Affairs, July 31, 2003. Mr. Winer, a former State Department official, is an independent consultant.

[66] Adel al-Jubeir, "U.S. and Saudi Officials Hold a News Conference on a Major Development in the War on Terrorism," Federal Document Clearing House Transcript, June 2, 2004.

[67] John J. Miller, "On the House...of Saud, that is: the Kingdom's Spending in America," *National Review*, 54, no. 6: 2002, p. 19.

[68] Adel al-Jubeir, "Saudi Plans to Fight Terrorism," Federal News Service, Dec. 3, 2002.

[69] It is estimated that there are 267 charities operating in Saudi Arabia. See Jamil al-Dhiyabi, "Supervisor of Saudi Charities to al-Hayat: There Are No Deception or Illegal Remittances in Our Work; Ministry To Apply New Measures Starting Next Year," *Al-Hayat* (London, in Arabic), October 6, 2003, translation by Foreign Broadcast Information Service (FBIS) online.

[70] Royal Embassy of the Kingdom of Saudi Arabia, "Saudi Arabian Monetary Agency Implements New Regulations Regarding Charities," June 12, 2003. Available at [http://www.saudiembassy.net/2003News/Press/PressDetail.asp?cYear=2003&cIndex=99].

[71] See Saudi Arabian Monetary Authority, "Rules Governing the Opening of Bank Accounts in Saudi Arabia and General Operational Guidelines," April 2003, Section 300-1-6-5. In June 2004, the Northern Virginia offices of the WAMY were raided by FBI and Customs officials in relation to a terrorism investigation. WAMY spokesmen have repeatedly denied the group is linked to terrorist organizations.

[72] *Agence France Presse*, "Saudi Arabia to Create Body for All Charity Abroad after Terror Charges," Feb. 28, 2004.

[73] Adel al-Jubeir, "U.S. and Saudi Officials Hold a News Conference on a Major Development in the War on Terrorism," Federal Document Clearing House Transcript, June 2, 2004.

[74] U.S. Department of the Treasury, "JS-1108: Treasury Announces Joint Action with Saudi Arabia Against Four Branches of al-Haramain In The Fight Against Terrorist Financing," January 22, 2004. Available at [http://www.treasury.gov/press/releases/js1108.htm].

[75] U.S. Department of the Treasury, "JS-1703: Additional al-Haramain Branches, Former Leader Designated by Treasury as Al Qaida Supporters," June 2, 2004. Available at [http://www.ustreas.gov/press/releases/js1703.htm].

[76] The designated branches are: Indonesia, Pakistan, the Netherlands, Afghanistan, Albania, Kenya, Tanzania, Bangladesh, Ethiopia, Bosnia, and Somalia.

[77] The designation of individuals and organizations under the regulations blocking or freezing terrorist assets requires an administrative determination by the Secretary of State or the Secretary of the Treasury, in consultation with one another and with the Attorney General. The initial determination is conducted without input from the subject of the determination and, thus, at this stage, the rigorous standards applicable to adversarial legal procedures such as criminal or civil trials are inapplicable.

[78] Majid Al-Kanani, "Al-Haramain Director Resigns," *Arab News* (Jeddah), July 15, 2004.

[79] This section of the report was prepared by Febe Armanios. For more details on *madrasa*s, see CRS Report RS21654, *Islamic Religious Schools, Madrasas: Background.* See also pages 1 and 10 of this report.

[80] Royal Embassy of the Kingdom of Saudi Arabia, "Statement by Adel Al Jubeir, Foreign Affairs Advisor to the Crown Prince," November

14, 2003, Available at [http://www.saudi embassy.net/press_release/releases/03-PR-1114-terrorism.htm].

[81] Available at [http://www.saudiembassy.net/ReportLink/SAMA%20INITIATIVES%20 BY%20 KSA%20UP%20DATED%20APRIL%202004.pdf].

[82] Patrick E. Tyler, "Stability Itself Is the Enemy," *New York Times*, Nov. 10, 2003.

[83] Susan Schmidt, "U.S. Officials Press Saudis on Aiding Terror," *Washington Post*, Aug. 6, 2003, p. A12.

[84] Testimony of Deputy Assistant Secretary of the Treasury Juan Zarate and Deputy Assistant Director of the FBI Counter-terrorism Operational Support Branch Thomas Harrington, Hearing of the Subcommittee on the Middle East and Central Asia of the House Committee on International Relations on Saudi Arabia and the Fight Against Terrorism Financing, March 24, 2004; Federal News Service transcript, pp. 9-10.

[85] *Ibid.*, Harrington, p. 10.

[86] Douglas Farah, "U.S.-Saudi Anti-Terror Operation Planned; Task Force Will Target Funding," *Washington Post*, Aug. 26, 2003.

[87] "U.S. and Saudis Join in Antiterror Effort," *New York Times*, Aug. 25, 2003; and Matthew Clark, "There's a New Task Force in Town," *The Christian Science Monitor*, Aug. 26, 2003.

[88] The Financial Action Task Force is "an inter-governmental body whose purpose is the development and promotion of policies, both at national and international levels, to combat money laundering and terrorist financing." For more see the FATF website at [http://www1.oecd.org/fatf/].

[89] See "Kingdom of Saudi Arabia: Executive Summary - FATF Recommendations for Anti-Money Laundering and Combating the Financing of Terrorism" in Annex C of the report Available at [http://www1.oecd.org/fatf/pdf/AR2004-Annexes_en.PDF]. The FATF's "40+8" Recommendations on terrorist financing and money laundering are internationally recognized regulatory and legal benchmarks.

[90] Saudi officials maintain that the wider body of Islamic *sharia* law that underlies the Saudi legal system adequately criminalizes material support for terrorism. U.S. and FATF officials argue that the language of Saudi Arabia's criminal statutes on terrorist financing should explicitly reflect international standards.

[91] Statement of David Aufhauser, Former General Counsel, Dept. of the Treasury, Middle East Policy Council forum, "Saudi Arabia: Enemy or Friend," Jan. 23, 2004. Available at [http://www.mepc.org /public_asp/forums_chcs/35.html].

[92] The implementing regulations for the Saudi anti-money laundering and terrorist financing law that was introduced in August 2003 have not been issued as of July 2004.

[93] Again, "multilateral organizations" based in Saudi Arabia like the IIRO, WAMY, and MWL are not subject to these restrictions.

[94] According to sources in the Treasury Department, Saudi Arabia is not yet a member of the Egmont Group, an international organization that links financial intelligence units (FIU) and facilitates information sharing. The Saudi Arabian FIU therefore does not yet have a formal relationship with the U.S. Financial Crimes Enforcement Network (FinCEN).

[95] "Show of Force," *Middle East Economic Digest*, 26 Mar.-1 Apr. 2004.

[96] U.S. Congress, House of Representatives, Hearing of the Subcommittee on the Middle East and Central Asia of the House Committee on International Relations on Saudi Arabia and the Fight Against Terrorism Financing, Rayburn House Office Building, Washington, DC, March 24, 2004; testimony of Mr. Thomas Harrington, Deputy Assistant director of the FBI Counterrorism Operational Support Branch, Federal News Service transcript, p. 7.

[97] *Ibid.*, testimony by (then) Deputy Assistant Secretary of the Treasury Juan Zarate, p. 8.

[98] Robin Allen, "Saudis meet anti-terror finance benchmarks," *Financial Times* (London), March 8, 2004.

[99] Available at [http://www.9-11commission.gov/staff_statements/911_ TerrFin_Monogr aph.pdf]

[100] Ibid.

[101] It is unclear how donations from the Saudi Popular Committee for Assisting the Palestinian Mujahideen to the PLO were considered in this regard.

In: Saudi Arabia: Terrorism, US Relations and Oil ISBN 1-59454-729-7
Editor: Nino P. Tollitz, pp. 63-87 © 2005 Nova Science Publishers, Inc.

Chapter 3

SAUDIA ARABIA: ENERGY AND ECONOMIC BACKGROUND*

Energy Information Administration

With one-fourth of the world's proven oil reserves and some of the lowest production costs, Saudi Arabia is likely to remain the world's largest net oil exporter for the foreseeable future. During January-October 2004, Saudi Arabia supplied the United States with 1.5 million barrels per day of crude oil, or 15%, of U.S. crude oil imports during that period.

GENERAL BACKGROUND

With oil export revenues making up around 90-95% of total Saudi export earnings, 70%-80% of state revenues, and around 40% of the country's gross domestic product (GDP), Saudi Arabia's economy remains, despite attempts at diversification, heavily dependent on oil (although investments in petrochemicals have increased the relative importance of the downstream petroleum sector in recent years).

Oil prices were strong during 2003, and even stronger in 2004. Combined with the highest Saudi oil output ever, the country's oil export revenues were up sharply in 2004, to more than $100 billion, compared to $77 billion in 2003 (and more than triple the $34 billion earned in 1998, when Saudi oil prices fell below $12 per barrel). For 2005 and 2006, oil export revenues are expected to remain very strong once again. Partly as a result of these strong oil export revenue increases, Saudi Arabia's real GDP growth was an estimated 6.1% in 2004, up from 1.7% average growth between 1995 and 2002. For 2005, Saudi real GDP growth is expected to remain strong, at 6.2%.

Meanwhile, during fiscal year 2004, Saudi Arabia originally had been expecting a budget deficit. However, this was based on an extremely conservative price assumption of $19 per barrel for Saudi oil -- and assumed production of 7.7 million barrels per day (bbl/d). Both of these estimates turned out to be far below actual levels. As a result, as of mid-December 2004, the Saudi Finance Ministry was expecting a huge budget surplus of $26.2 billion, on budget revenues of $104.8 billion (nearly double the country's original estimate) and expenditures of $78.6 billion (28% above the approved budget levels). This surplus is being used for several purposes,

* Excerpted from http://www.eia.doe.gov/emeu/cabs/saudi.html

including: paying down the Kingdom's public debt (to $164 billion from $176 billion at the start of 2004); extra spending on education and development projects; increased security costs (possibly an additional $2.5 billion dollars in 2004; see below) due to threats from terrorists; and higher payments to Saudi citizens through subsidies and other means. For 2005, Saudi Arabia is assuming a balanced budget, with revenues and expenditures of $74.6 billion each.

In spite of the recent surge in its oil income, Saudi Arabia continues to face serious long-term economic challenges, including high rates of unemployment (14% or greater), one of the world's fastest population growth rates (2.4% per year), and the consequent need for increased government spending. Saudi Arabia's per capita oil export revenues (in constant, $2004) remain far below high levels reached during the 1970s and early 1980s (around $4,511 per person in 2004 versus $22,174 in 1980, for instance). This is in large part due to the fact that Saudi Arabia's young population has nearly tripled since 1980, while oil export revenues in real terms have fallen sharply (also, in recent years the value of the dollar -- the currency in which oil is traded -- has dropped, hurting Saudi Arabia's "terms of trade"). Meanwhile, Saudi Arabia has faced nearly two decades of heavy budget and trade deficits, the expensive 1990/1991 war with Iraq, and total government debt approaching three-quarters of Saudi GDP. On the other hand, Saudi Arabia does have extensive foreign assets, including $23 billion in foreign exchange reserves and gold, which provide a substantial fiscal "cushion." Finally, surging oil export revenues remove an incentive for Saudi Arabia to undertake needed economic reforms, such as reduction in the bloated state sector, as urged by the IMF and others.

Movement towards economic reform in Saudi Arabia remains uneven at best. For instance, reducing subsidies and increasing tariffs on electricity have proven problematic, with a rate increase announced in April 2000 subsequently reversed in October in the face of widespread public opposition. Besides economic reform, Saudi Arabia also has made only slow progress on another of its main domestic goals -- attracting foreign direct investment (FDI). In January 2004, the Saudi cabinet approved a reduction in taxes on foreign direct investment (to 20% in most sectors; 30% in the natural gas sector) as part of an effort to speed up the economic reform and privatization process in the country.

Currently, large state corporations, like oil firm Saudi Aramco (which has a monopoly on Saudi upstream oil development and controls 98% of the country's oil reserves) and the Saudi Basic Industries Corporation (SABIC; the world's 11th largest petrochemical producer) dominate the Saudi economy. To date, there has not been a single sale of state assets to private control, and privatization largely has been limited to allowing private firms to take on certain service functions. In May 2002, Saudi Oil Minister Ali Naimi (reappointed in May 2003 for a third, four-year term) stated that the country was considering privatizing some operations of Saudi Aramco. One impetus for Saudi privatization is its desire to join the World Trade Organization (WTO), but progress has been slow towards achieving this goal, and there were no signs of an imminent breakthrough as of December 2004.

In general, Saudi Arabia also has moved cautiously and slowly towards government subsidy cuts, tax increases, or financial sector reforms. Saudi leadership (Crown Prince Abdullah, in particular) has indicated that it sees privatization -- although controversial -- as a "strategic choice," and has created (in August 1999) a "Supreme Economic Council" charged with boosting investment, creating jobs for Saudi nationals, and promoting privatization. Changes to the law governing foreign investment, granting the same basic rights to foreign investors as to Saudi nationals, were approved in April 2000. Among other measures, the new investment law reduced taxes on foreign business profits from a maximum 45% to 30%, and provided greater legal protections against expropriation of investments.

In November 1999, King Fahd stated that "the world is heading for...globalization" and that "it is no longer possible for [Saudi Arabia] to make slow progress." In the context of successfully becoming integrated into the global economy, Fahd also emphasized the importance of regional unity among Gulf states -- economically, politically, and militarily. Along these lines, a customs union among GCC countries was agreed upon at the December 1999 GCC summit. The union is to take effect in March 2005. Currently, goods from GCC countries are exempt from all Saudi import duties, as long as 40% of their value has been added within the GCC and the producing company is owned at least 51% by GCC citizens. The GCC has also agreed to impose a common set of Value Added Taxes (VATs) beginning in 2005.

In a treaty signed in June 2000, Saudi Arabia and Yemen agreed on the delineation of sections of their common border which had been in dispute since the 1930s. The deal is expected to open up opportunities for increased Saudi trade and investment in Yemen, a possible pipeline across Yemen to the Arabian Sea (see below for more details), and the possible award of oil and gas exploration rights for areas in Yemen adjacent to previously disputed areas of the border. In February 2001, Saudi Arabia and Syria signed a bilateral free-trade agreement. On June 11, 2001, Saudi Arabia announced (in a letter to UN Secretary General Kofi Annan) that it was taking ownership of Iraq's pipeline to the Saudi Red Sea coast (closed since August 1990), citing Iraqi threats and aggressive actions, including (allegedly) a series of cross-border raids in recent months. Saudi Arabia said that Iraq's behavior had "destroyed any rationale for maintaining the [pipeline] facilities."

In May 2000, a new law aimed at attracting foreign investment to the Saudi energy sector came into effect. The law permits full foreign ownership of Saudi property and licensed projects, sets up the General Investment Authority (SAGIA) as a "one-stop shop" for foreign investors, and reduces taxes on company profits from 45% to 30%. Previously, foreign companies were limited to a 49% share of joint ventures with Saudi domestic partners. Several important sectors, however, remain closed to 100% foreign ownership, including (as of June 2003): upstream oil, pipelines, media and publishing, insurance, telecommunications, defense and security, health services, wholesale and retail trade, and more. Thus, the new foreign investment law is far less attractive than it appears at first glance. However, in January 2001, SAGIA reported that foreign investment commitments had reached $1.6 billion, including 53 licenses (29 industrial, 24 non-industrial)

OIL

According to the Oil and Gas Journal, Saudi Arabia contains 261.9 billion barrels of proven oil reserves (including 2.5 billion barrels in the Saudi-Kuwaiti Divided, aka "Neutral" Zone), around one-fourth of proven, conventional world oil reserves. Around two-thirds of Saudi reserves are considered "light" or "extral light" grades of oil, with the rest either "medium" or "heavy." Overall, Saudi Arabia may contain up to 1 trillion barrels of ultimately recoverable oil, with Oil Minister Naimi stating on December 27, 2004 that the country's proven reserves could reach 461 billion barrels within a few years. Although Saudi Arabia has around 80 oil and gas fields (and over 1,000 wells), more than half of its oil reserves are contained in only eight fields, including Ghawar (the world's largest oil field, with estimated remaining reserves of 70 billion barrels) and Safaniya (the world's largest offshore oilfield, with estimated reserves of 35 billion barrels). Ghawar's main producing structures are, from north to south: Ain Dar, Shedgum, Uthmaniyah, Farzan, Ghawar, Al Udayliyah, Hawiyah, and Haradh. Overall, Ghawar alone accounts for about half of Saudi Arabia's total oil production capacity.

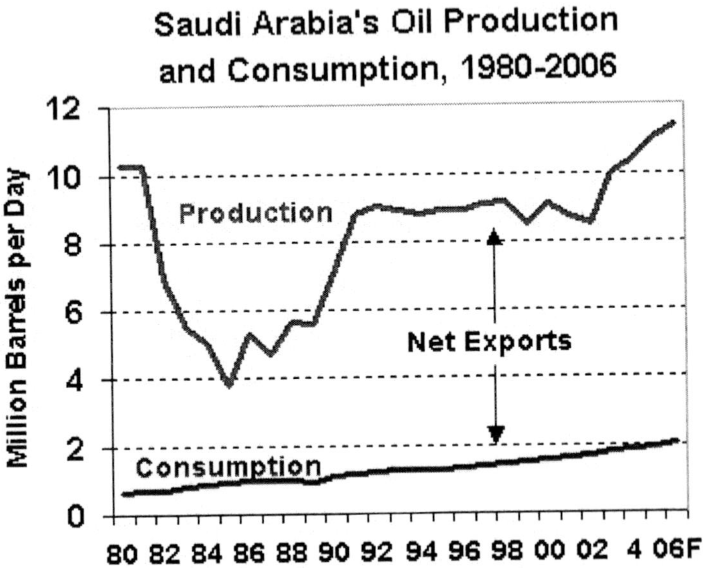

Note: Production includes crude oil, natural gas liquids, other liquids, and refinery gain.

Saudi Arabia is the world's leading oil producer and exporter, and its location in the politically volatile Gulf region adds an element of concern for its major customers, including the United States. Saudi Arabia maintains crude oil production capacity of around 10.5-11.0 million bbl/d, and claims that it is "easily capable" of producing up to 15 million bbl/d in the future and maintaining that production level for 50 years. The Saudis also have stated recently that they intend to raise production capacity to 12.5 million bbl/d over the next several years. One challenge for the Saudis in achieving this objective is that their existing fields sustain 5%-12% annual "decline rates," (according to Aramco Senior Vice President Abdullah Saif, as reported in Petroleum Intelligence Weekly and the International Oil Daily) meaning that the country needs to add 600,000-800,000 bbl/d in new capacity each year just to compensate.

Aramco claims that the average total depletion for Saudi oil fields is 28%, with the giant Ghawar field having produced 48% of its proved reserves. Aramco also claims that, if anything, Saudi oil reserves are underestimated, not overestimated. Some outside analysts, notably Matthew Simmons of Houston-based Simmons and Company International, have disputed Aramco's optimistic assessments of Saudi oil reserves and future production, pointing to -- among other things -- more rapid depletion rates and a higher "water cut" than the Saudis claim. EIA forecasts that Saudi oil production capacity could reach 18.2 million bbl/d by 2020, and 22.5 million bbl/d by 2025, a prediction which the Saudis have said is "unrealistic." In recent months, Saudi Arabia reportedly has stepped up oil drilling and exploration activity, aiming to maintain a total of 60 rigs -- double the number as of December 2004.

PRODUCTION

During 2004, Saudi Arabia produced around 10.4 million bbl/d of total oil -- including crude oil, natural gas liquids, and "other liquids"oil, and also including half of the Saudi-Kuwaiti Divided Zone's 610,000 bbl/d). This was up sharply from Saudi Arabia's 8.5 million bbl/d of total oil production in 2002 (see graph). Saudi oil production surged in 2003 as the country utilized its spare production capacity to help compensate for losses from Venezuela, Nigeria, and Iraq. Also, in early June 2004, OPEC met and agreed to raise crude oil output quotas by 2 million bbl/d as part of an effort towards lowering oil prices, which had surpassed $40 per barrel. As a result, beginning on July 1, 2004, Saudi Arabia's official OPEC crude oil production quota was raised to 8.288 million bbl/d, up from the country's previous quota of 7.638 million bbl/d. Currently (late December 2004), Saudi Arabia is estimated to be producing 9.4 million bbl/d of crude oil, well in excess of its current quota level of 8.775 million bbl/d (effective November 1, 2004). In addition to crude oil, Saudi Arabia produces around 1.3 million bbl/d of natural gas liquids (NGLs) and "other liquids," not subject to OPEC quotas.

Saudi Arabia produces a range of crude oils, from heavy to super light. Of Saudi Arabia's total oil production capacity, about 65%-70% is considered light gravity, with the rest either medium or heavy; the country is moving to reduce the share of the latter two grades. Lighter grades generally are produced onshore, while medium and heavy grades come mainly from offshore fields. The Ghawar field is the main producer of 34o API Arabian Light crude, while Abqaiq (a super-giant field with 17 billion barrels of proven reserves) produces 37o API Arab Extra Light crude. Since 1994, the Hawtah Trend (also called the Najd fields), which includes the Hawtah field and smaller satellites (Nuayyim, Hazmiyah) south of Riyadh, has been producing around 200,000 bbl/d of 45o-50o API, 0.06% sulphur, Arab Super

Light. Overall, the Najd fields are estimated to contain 30 billion barrels of liquids and major reserves of natural gas. Offshore production includes Arab Medium crude from the Zuluf (over 500,000 bbl/d capacity) and Marjan (270,000 bbl/d capacity) fields and Arab Heavy crude from the Safaniya field. Most Saudi oil production is considered "sour," containing relatively high levels of sulfur.

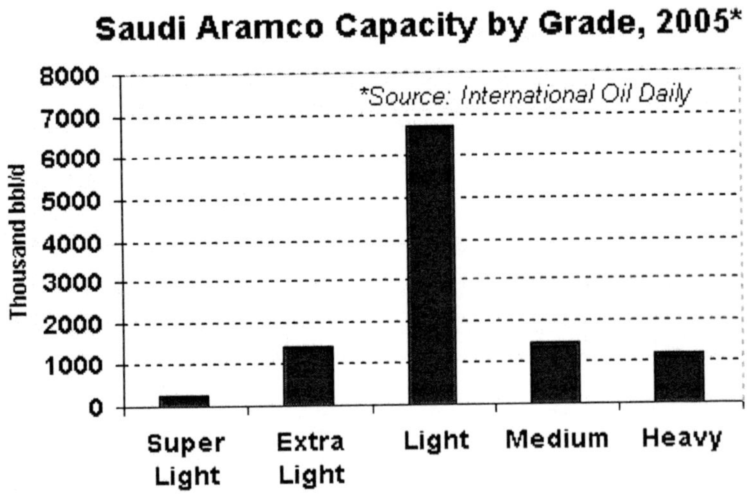

Saudi Arabia's long-term goal is to develop its lighter crude reserves, including the Shaybah field located in the remote Empty Quarter area bordering the United Arab Emirates. Shaybah contains an estimated 15.7 billion barrels (or higher) of premium grade 41.6o API sweet (nearly sulfur-free) crude oil, with production potential believed to be 1 million bbl/d (output as of early April 2004 was around 560,000 bbl/d). Also, in October 2004, Aramco announced that it had discovered 54o API super-light crude oil in the Abu Sidr region, 115 miles south of Riyadh.

Overall, the Shaybah project cost around $2.5 billion, with production starting in July 1998. According to Oil Minister Naimi (October 1999), the development of Shaybah showed that "the cost of adding...capacity - that is, all the infrastructure, producing and transportation facilities - necessary to produce one additional barrel of oil per day in Saudi Arabia is, at most, $5,000 compared to between $10,000 and $20,000 in most areas of the world. So both our current production costs, and the costs for developing more production capacity for the future, are probably the world's lowest."

The Shaybah complex includes three gas/oil separation plants (GOSPs) and a 395-mile pipeline to connect the field to Abqaiq, Saudi Arabia's closest gathering center, for blending with Arab Light crude (Berri and Abqaiq streams). In addition to oil, Shaybah has a large natural gas "cap" (associated gas), with estimated reserves of 25 trillion cubic feet (Tcf). Gas production of 880 million cubic feet per day (Mmcf/d) is reinjected, along with natural gas liquids (NGLs). A possible gas recovery project could be implemented within 5 or 6 years, potentially for use in petrochemical production.

In March 2002, Aramco awarded major turnkey contracts to Italy's Snamprogetti ($630 million) and Technip-Coflexip ($360 million) aimed at increasing total Saudi oil production capacity by 800,000 bbl/d (500,000 bbl/d of Arabian light and 300,000 bbl/d of Arabian medium), by late 2004. The $1.2 billion project, known as the Qatif producing facilities development program (QPFDP), is located in the eastern part of the country near Dhahran. QPFDP involves construction of two gas-oil separation plants (GOSPs), as well as gas treatment and oil stabilization facilities, for the Qatif and Abu Safa oilfields. Additional Qatif and Abu Safa production had been slated to replace production elsewhere in Saudi Arabia, not to boost overall capacity, although recently this issue has been thrown into some question as the Saudis attempt to maintain a spare capacity cushion in the face of rapidly growing world oil demand. As of December 2004, Saudi Arabia reportedly had brought production from Qatif and Abu Safa online.

Another potential project, at the Khurais field (a satellite of Ghawar), could increase Saudi production capacity by 800,000 bbl/d at a cost of $3 billion. This would involve installation of four GOSPs, with a capacity of 200,000 bbl/d each, at Khurais, which first came online in the 1960s but was mothballed by Aramco.

Several other fields -- Abu Hadriya (1.8-2.0 billion barrels in reserves), Fadhili (1-1.4 billion barrels), Harmaliyah, Khursaniyah (4 billion barrels), and Manifa -- were mothballed by the Saudis during the the 1990s, but could be brought back online given high world oil demand and the desire to maintain Saudi spare production capacity. In particular, Saudi Aramco appears to be pushing ahead with development of the Abu Hadriya, Fadhili and Khursaniya (AFK) onshore fields. Production of 500,000 bbl/d (medium, 35o API) from the AFK fields is slated to begin in late 2007. Besides AFK, the Saudis are considering development of the offshore Manifa field in coming years.

The $280 million Haradh-2 project aims to increase production capacity at the Haradh oil field to 900,000 bbl/d -- triple current production -- by 2006. This will involve adding a second, 300,000-bbl/d GOSP to Haradh (in addition to one 300,000-bbl/d GOSP inaugurated in January 2004), while shutting in some heavy oil production in the Ghawar area. Haradh also will produce significant volumes of non-associated natural gas, natural gas condensates (perhaps 170,000 bbl/d), and sulfur. The project is being carried out by Aramco, along with private companies like Foster-Wheeler.

SAUDI-KUWAITI DIVIDED ZONE; BAHRAIN

The Saudi-Kuwait Divided Zone contains about 5 billion barrels of proven oil reserves. Within the Divided Zone, Japan's Arabian Oil Co. (AOC) traditionally had operated two offshore fields (Khafji and Hout) with 300,000 bbl/d in production, but in February 2000, it lost the concession (in January 2003, AOC reached an agreement with Kuwait on the right to purchase at least 100,000 bbb/d of crude for the next 20 years from Khafji). The offshore Saudi Divided Zone had represented Japan's most significant upstream oil interest, with 80% of revenues going to AOC and 10% each to Saudi Arabia and Kuwait. ChevronTexaco, meanwhile, operates three onshore fields (Wafra, South Fawaris, and South Umm Gudair) in the Divided Zone. Saudi Arabia had stated that it wanted AOC and Japan to increase their investments in Saudi Arabia (including more than $1 billion in a railway linking remote mining areas to export terminals), as well as their purchases of Saudi oil, as a condition for renewal of AOC's drilling rights in the Divided Zone. In August 2001, Saudi Arabia rejected a request by AOC to reopen talks on the Saudi Divided Zone concession. In September 2002, Saudi Oil Minister Naimi proposed that Japan invest in Saudi natural gas fields rather than oil.

Besides the Kuwaiti-Saudi Divided Zone, Saudi Arabia also produces oil jointly with Bahrain, from the Abu Safa offshore oilfield. In addition to what is produced in its territory, Bahrain receives oil via subsea pipeline from Abu Safa. As a way of supporting their neighbor's economy, the field's Saudi administrators have donated all of the income from its 150,000 bbl/d of production to Bahrain (as of late 2004, with output from Abu Safa doubling to 300,000 bbl/d, the Saudis apparently reduced this share to 50%). The remainder of the oil that Bahrain receives from Saudi Arabia comes from the oil fields around Dammam. The Abu Safa pipeline passes through this area on its way to Bahrain. Unlike the Abu Safa oil which it receives as a gift, Bahrain purchases the Dammam oil, albeit at a discounted rate (as of late 2004, there were indications that the Saudis might stop providing Dammam oil to Bahrain as a gift).

EXPORTS, PORTS, PIPELINES, SHIPPING

Saudi Arabia is a key oil supplier to the United States and Europe. Asia (e.g., China, Japan, South Korea, India) now takes around 60% of Saudi Arabia's crude oil exports, as well as the majority of its refined petroleum product exports. During the first ten months of 2004, Saudi Arabia exported 1.55 million bbl/d of oil (of which 1.49 million bbl/d was crude) to the United States. For this time period, Saudi Arabia ranked third (after Canada and Mexico, and just ahead of Venezuela) as a source of total (crude plus refined products) U.S. oil imports, and third for crude only. Saudi Arabia is eager to maintain and even expand its market share in the United States for a variety of economic and strategic reasons. During the first ten months of 2004, Saudi Arabia's share of U.S. crude oil imports was 14.8%, down from 18.2% during the first ten months of 2003.

Most of Saudi Arabia's crude oil is exported from the Persian Gulf via the huge Abqaiq processing facility, which handles around two-thirds or so of the country's oil output. Saudi Arabia's primary oil export terminals are located at Ras Tanura (6 million bbl/d capacity; the world's largest offshore oil loading facility) and Ras al-Ju'aymah (3 million bbl/d) on the Persian Gulf, plus Yanbu (as high as 5 million bbl/d) on the Red Sea. Combined, these terminals appear capable of handling around 14 million bbl/d, around 3.0-3.5 million bbl/d higher than Saudi crude oil production capacity (10.5-11.0 million bbl/d), and about 6 million bbl/d in excess of Saudi crude oil production in 2002. Despite this excess capacity, there have been reports that the Saudis are planning to conduct a feasibility study on construction of an oil pipeline from the Empty Quarter of southeastern Saudi Arabia through the Hadramaut in Yemen to the Arabian Sea.

Saudi Arabia operates two major oil pipelines. The 5-million-bbl/d East-West Crude Oil Pipeline (Petroline), operated by Aramco since 1984 (when it took over from Mobil), is used mainly to transport Arabian Light and Super Light to refineries in the Western Province and to Red Sea terminals for export to European markets. The Petroline was constructed in 1981, with initial capacity of 1.85 million bbl/d on a single, 48-inch line (AY-1). The Petroline was expanded in 1987, during the height of the Iran-Iraq war (and specifically the so-called "tanker war" in the Gulf), to 3.2 million bbl/d, with the addition of a parallel ("looped") , 56-inch line (AY-1L). Finally, in 1993, Petroline capacity was increased to 5.0 million bbl/d by adding significant pumping capability on the line. Reportedly, the Saudis expanded the Petroline in part to maintain Yanbu as a strategic option to Gulf port facilities in the event that exports were blocked at that end. A study in 1997 by the Baker Institute indicated that capacity on the Petroline could be expanded significantly by using so-called "drag reduction agents" (DRAs), and that this could enhance the line's strategic value.

In purely economic terms, Yanbu remains a far less economical option for Saudi oil exports than Ras Tanura. Among other factors, shipments from Yanbu add about 5 days roundtrip travel time for tankers through the Bab al-Mandab strait to major customers in Asia compared to Ras Tanura (via the Strait of Hormuz). In addition, according to Oil Minister Naimi, the Petroline is only utilized at half capacity. Given this fact, as well as the desire to boost natural gas usage (see below), Saudi Aramco has begun converting the 48-inch (AY-1) line to natural gas pumping capability. The natural gas will supply Yanbu's petrochemical and power facilities.

Running parallel to the Petroline is the 290,000-bbl/d Abqaiq-Yanbu natural gas liquids pipeline, which serves Yanbu's petrochemical plants. The Trans-Arabian Pipeline (Tapline) to Lebanon is mothballed, and the 1.65-million-bbl/d, 48-inch Iraqi Pipeline across Saudi Arabia (IPSA), which runs parallel to the Petroline from pump station #3 (there are 11 pumping stations along the Petroline, all utilizing on-site gas turbine electric generators) to the port of Mu'ajjiz, just south of Yanbu, was closed indefinitely following the August 1990 Iraqi invasion of Kuwait. In June 2001, Saudi Arabia seized ownership of IPSA "in light of the Iraqi government's persistence in its stands." Theoretically, IPSA could be used for Saudi oil transport to the Red Sea, although the Saudis have stated that "there are no plans" to do so. According to Oil Minister Naimi, Saudi Arabia has "surplus oil export and pipelines capacity...[including the] East-West oil pipeline system [which] can carry and deliver 5 million bbl/d" but is being run at "only half capacity."

Aramco's shipping subsidiary Vela has the world's largest fleet of oil tankers, including 19 VLCC's (very large crude carriers) and 4 ULCC's (ultra large crude carriers). Overall, Vela carries around half of Saudi oil exports. In September 2004, the Saudis placed a $200 million for two VLCCs from Hyundai Heavy Industry, with delivery expected in 2007. In addition to tankers, Aramco owns or leases oil storage facilities around the world, in places like Rotterdam, Sidi Kerir (the Sumed pipeline terminal on Egypt's Mediterranean coast), South Korea, the Philippines, the Caribbean, and the United States.

REFINING AND PETROCHEMICALS

Saudi Arabia has eight refineries, with combined crude throughput capacity of around 1.75 million bbl/d, plus around 1.6 million bbl/d of refining capacity overseas. A new, 200,000-bbl/d fractionation unit at the Ras Tanura refinery was completed in August 2003, raising Saudi refining capacity close to 2 million bbl/d. Also slated for upgrading is the Rabigh refinery on the Red Sea coast. Plans call for boosting capacity at Rabigh, Saudi Arabia's largest domestic refinery, to as high as 400,000 bbl/d, as well as upgrading the refinery's product slate away from low-value heavy products towards gasoline and kerosene at an estimated cost of $3-$4 billion. In addition, the project will add a world-class ethane cracker fed by natural gas from the Eastern Province via a converted oil line of the East-West crude oil pipeline. In May 2004, Japan's Sumitomo agreed to a $4.3 billion joint venture deal to build a major petrochemical plant at Rabigh. The plant is to come onstream in 2008, and will produce ethylene and propylene.

In July 2004, Aramco signed an agreement with Shell to purchase a 9.96% share in Showa Shell Group, a refining and marketing company based in Japan. Under the deal, Aramco will supply Showa Shell with 300,000 bbl/d of crude oil.

Saudi Arabia has ambitious plans for expanding petrochemical production using natural gas as a feedstock. State-owned (70%) SABIC, the Middle East's largest non-oil industrial company (and expected to become one of the world's top five ethylene producers by 2005), accounts for around 10% of world petrochemical production. In February 2001, SABIC completed a $1 billion expansion at the Yanbu petrochemical facility, making it the largest polyethylene plant in the world, and in April 2003, Aramco awarded a turnkey contract to Snamprogetti to build to new units at the complex. In early January 2002, SABIC agreed to a $1.15 billion loan to fund a new petrochemicals plant in the eastern Saudi Arabian industrial city of Jubail. The complex is scheduled to come online in the second half of 2004, and to produce 1 million tons per year of ethylene, plus olefins, polyethylene, and glycol ethylene.

SECURITY ISSUES

In December 2004, an attack on the U.S. consulate in Jeddah killed five staff members. In the same month, two car bombs exploded in Riyadh, wounding several people. During the first two weeks of June 2004, several Westerners were murdered, including one who was kidnapped and beheaded. Even before these recent attacks, several terrorist incidents in April and May 2004 (at ABB Lummus offices in Yanbu, security headquarters in Riyadh, and a residential compound in Khobar) had raised concerns about security in the Kingdom. Specifically, these concerns centered around Saudi oil facilities as well as the 3,000 Western oil workers in the Kingdom. Meanwhile, in mid-December 2004, a Saudi branch of al-Qaeda posted a message on its website urging its members to "strike all foreign targets and the hideouts of the tyrants to rid the peninsula of the infidels and their supporters."

In late April 2004, Aramco's Chief Executive, Abdullah Jumah, said that "there is nowhere in the world that oil facilities are protected as well as in Saudi Arabia." According to Jumah, Aramco employs 5,000 security guards to protect oil facilities. In addition, the Saudi National Guard, regular Saudi military forces, and Interior Ministry officers are tasked with protecting oilfields, pipelines (the country has around 10,000 miles), ports (Ras Tanura, Al Juaymah, Yanbu), refineries, and other oil facilities (gathering centers, gas-oil separation plants, etc.). In May 2004, Nawaf Obaid, an advisor to the Saudi royal family, said that the Saudi government had added $750 million to its security budget over the past two years to beef up security in the oil sector. According to Obaid, the Saudis spent $5.5 billion in 2003 on oil security. In addition to direct security, Saudi Arabia is known to maintain "redundancy" (i.e., multiple options for transportation and export) in its oil system, in part as a form of indirect security against any one facility being disabled.

NATURAL GAS

According to Oil and Gas Journal, Saudi Arabia's proven natural gas reserves are estimated at 235.0 trillion cubic feet (Tcf), ranking fourth in the world (after Russia, Iran, and Qatar), and up about 5 Tcf from 2002. Most (around 60%) of Saudi Arabia's currently proven natural gas reserves consist of associated gas, mainly from the onshore Ghawar field and the offshore Safaniya and Zuluf fields. The Ghawar oil field alone accounts for one-third of the country's proven natural gas reserves. However, it is important to note that only 15% of Saudi Arabia has been "adequately explored for gas," according to Aramco's vice president for new business development, Khalid al-Falih. Most new associated natural gas reserves discovered in the 1990s have been in fields which contain light crude oil, especially in the Najd region south of Riyadh. Most of Saudi Arabia's non-associated gas reserves (Mazalij, Al-Manjoura, Shaden, Niban, Tinat, Al-Waar, etc.) are located in the deep Khuff reservoir, which underlies the Ghawar oil field. Natural gas also is located in the countries extreme northwest, at Midyan, and in the Empty Quarter (Rub al Khali) in the country's southeastern desert. The Rub al Khali alone is believed to contain natural gas reserves as high as 300 Tcf.

Another large natural gas field, called Dorra, is located offshore near the Khafji oil field in the Saudi-Kuwaiti Divided Zone and may be developed by Japan's AOC. Dorra development is controversial, however, because part of it is also claimed by Iran (which calls the field Arash). The maritime border between Kuwait and Iran remain undemarcated, but Saudi Arabia reached an agreement with Kuwait in July 2000 to share Dorra equally. Currently, Iran is resisting any moves by Kuwait and Saudi Arabia to develop the field on their own.

In June 2003, Saudi Oil Minister Naimi officially announced termination of negotiations with foreign energy companies on the $15-$20 billion "Saudi Gas Initiative" (SGI), which had promised to be the first major reopening of Saudi Arabia's upstream hydrocarbons sector to foreign investment since nationalization in the 1970s. Companies which had been selected (in 2001) for the three "core ventures" under the SGI were: 1) South Ghawar -- ExxonMobil (35%), Shell (25%), BP (25%), Phillips (15%); 2) Red Sea -- ExxonMobil (60%), plus Marathon (20%) and Occidental (20%); and 3) Shaybah -- Shell (40%), Total (30%), and Conoco (30%). The SGI had aimed to increase foreign investment and natural gas development in the country, while integrating upstream gas development with downstream petrochemicals, power generation, and water desalination. SGI had been seen as the key to Saudi Arabia's entire foreign investment strategy. However, negotiations broke down over two major stumbling blocks: the extent of gas reserves to be opened to upstream development and whether or not this should include gas from the Saudi Aramco Reserve Area (SARA); and the rates of return to participating companies (the companies wanted a significantly higher rate than the Saudis were offering).

Core Venture 1, in South Ghawar, would have been one of the world's largest ($15 billion) integrated natural gas projects, including exploration, pipelines, two gas-fired power plants, two petrochemical plants, two desalination units, and more. Core Venture 2 was to involve exploration in the Red Sea, development of the Barqan and Midyan fields on the Red Sea coast in northwestern Saudi Arabia, as well as construction of a petrochemical plant, a power station, desalination capacity, etc., at a cost of

$4 billion. Core Venture 3 would have involved exploration near Shaybah in the Empty Quarter, development of the Kidan gas field, laying of pipelines from Shaybah to the Haradh and Hawiyah natural gas treatment plants east of Riyadh, and construction of a petrochemical plant in Jubail, at a cost of $4 billion.

Following cancellation of the SGI, Saudi Arabia repackaged the project as a series of smaller, more focused contracts, with better rates of return than previously offered. At the same time, the Saudis moved away from the integrated upstream/downstream gas, water, power, and petrochemical nature of the SGI, and instead specifically targeted upstream natural gas development in the area that had comprised Core Venture 3. Downstream and "midstream" elements of the SGI will now be handled separately, in large part by SABIC and Aramco. In July 2003, Saudi Arabia reached a tentative deal (officially signed on November 15) with Royal Dutch/Shell and Total on Blocks 5-9 and 82-85 in the Shaybah and Kidan areas of the Empty Quarter region. Besides the major European companies, Saudi Aramco -- replacing ConocoPhillips -- will have a 30% share in the $2 billion project. Shell will maintain a 40% share and Total the remaining 30%, in a consortium known as the South Rub al-Khali Company (SRAK). The deal covers an area of 81,000 square miles.

In January 2004, Russia's Lukoil won a tender to explore for and produce non-associated natural gas in the Saudi Empty Quarter. Lukoil will operate in Block A, near Ghawar, as part of an 80/20 joint venture (called "Luksar") with Saudi Aramco. Also in January 2004, China's Sinopec won a tender for gas exploration and production in Block B, while an Eni-Repsol consortium was granted a license to operate in Block C. Under terms of the agreements, Aramco will take "sales quality gas" on a take-or-pay basis for $0.75 per million Btu, while condensates and natural gas liquids will be sold at international market rates (note: Saudi accession to the WTO will most likely require it to give up the dual pricing system for natural gas, and also to set up a comprehensive, transparent regulatory framework for the natural gas sector). In addition, the Saudi government will fund a pipeline connection from the country's Master Gas System (MGS) to contract delivery points.

In October 2002, construction was completed on a $4 billion, 1.4-billion-cubic-feet (Bcf)-per-day, non-associated gas processing plant at Hawiyah, located south of Dhahran and east of Riyadh near the giant Ghawar oil field. Hawiyah represents the largest Saudi natural gas project in more than 10 years, and the first to process only non-associated gas (from the deep Khuff and Jauf reservoirs). Hawiyah was officially inaugurated in October 2002, and reportedly is producing enough natural gas to free up around 260,000 bbl/d of Arabian Light crude oil for export. Aramco also has invited bids to expand Hawiyah to recover "hundreds of thousands of barrels daily of additional petrochemical feedstock," primarily NGLs from the treatment of 4 billion cubic feet (Bcf) per day of natural gas.

Besides Hawiyah, Foster Wheeler has been managing a $2 billion project to build a new natural gas processing plant at Haradh, 120 miles southwest of Dhahran at the southern tip of Ghawar. The Haradh plant was completed in the summer of 2004, increasing total Saudi natural gas processing capability by 1.6 Bcf/day, to around 9.5 Bcf/day. Haradh processes non-associated natural gas (both sweet and sour) from four fields in the Khuff formation. In addition, a $1.2 billion, 3,800-Mmcf/d "straddle plant" -- a natural gas reprocessing plant located adjacent to a gas transmission line for the purpose of extracting light hydrocarbon liquids newly formed due to recurring compression and decompression of gas during transmission -- is slated to be built. When complete, the straddle plant will service both Haradh and Hawiyah and increase Saudi NGL production.

In other natural gas-related developments, a key pipeline project was completed in June 2000 to extend the MGS from the Eastern Province (which contains large potential gas and condensate reserves) to the capital, Riyadh, in the Central Province. This is part of a broader expansion of the existing gas transmission system in Saudi Arabia, reportedly to include the construction of around 1,200 miles of additional natural gas pipeline capacity (on top of 10,500 miles of oil, gas, condensate, products, and natural gas liquid pipelines currently in operation) by 2006.

Domestic demand is driving expansion of the MGS, which was completed in 1984. The MGS feeds gas to the industrial cities of Yanbu on the Red Sea and Jubail, which combined account for 10% of the world's petrochemical production. Prior to the MGS, all of Saudi Arabia's natural gas output was flared.

ELECTRICITY

Saudi Arabia's rapidly growing population and artificially low power prices (as a result of low, government mandated tariffs and consumer subsidies) are increasing demand on electric utilities, as power demand grows by 7% or more each year (see graph). Saudi Arabia's Industry and Electricity Ministry estimates that the country will require up to 20 gigawatts (GW) of additional power generating capacity by 2019 -- nearly the same amount as today's 26.6 GW -- at a cost of $4.5-$6 billion per year. Most of this money is slated to come from the private sector, possibly including foreign investors. Also, the vast majority of this capacity will either be natural gas-fired or combined cycle, as part of the government's plans to expand gas utilization in the power sector (and elsewhere) significantly.

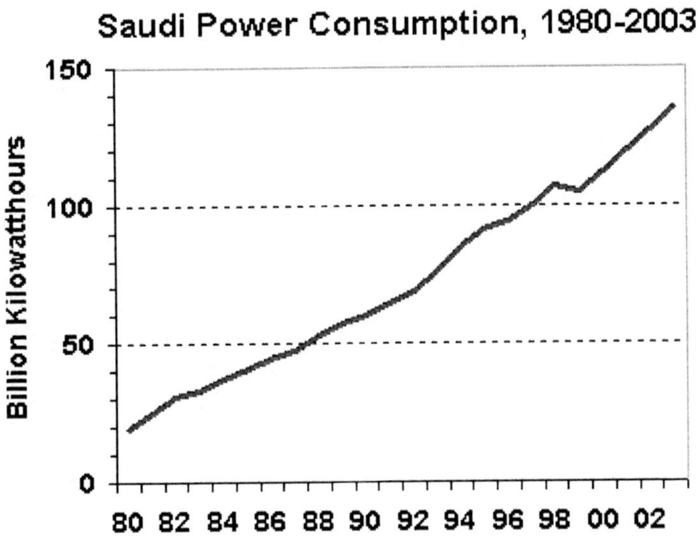

On February 16, 2000, Electricity Minister Dr. Hashem Ibn Abdullah Yamani signed a merger agreement between Saudi Arabia's 10 existing regional power companies (SCECOs), and on April 5, 2000, the long-anticipated SEC, a joint-stock company owned 50% by the Saudi government, was established. Creation of the SEC could open the door to private sector construction of new power plants on BOO (Build-Own-Operate) and BOT (Build-Own-Transfer) bases. The future of IPP's (Independent Power Producers) in Saudi Arabia remains uncertain, however, with major challenges including tariffs, the legal and operating framework, taxation, and fuel supply. In January 2003, the Electricity Services Regulatory Authority (ESRA) was set up as an independent "watchdog" in charge of the country's power sector, IPPs, and IWPPs.

Several recent projects have employed financing mechanisms that are new to Saudi Arabia's electric power sector. For example, the $1.7 billion, Ghazlan II power project was financed by an internationally syndicated, $500 million, commercial loan (the first such loan in Saudi history), and was built by a consortium led by Mitsubishi and Bechtel. Ghazlan II consists of four, 600-MW steam turbine units, the first of which came online in the summer of 2001. Combined with the existing 1,600-MW Ghazlan I facility located on the Gulf coast north of Dammam, the entire complex has a power generating capacity of 4,000 MW and supplies Saudi Arabia's Eastern Province.

In July 2002, the Supreme Economic Council passed a resolution setting out a framework for private sector involvement in developing independent water and power projects (IWPPs). Saudi Arabia reportedly is hoping to attract private sector investment for up to 60% equity in IWPP projects. Initial IWPP projects identified for development include a $1.4-$1.6 billion, 650-900-MW, 194-million-gallons-per-day (mmg/d) oil-fired plant at Shuaiba on the Red Sea coast 70 miles southeast of Jiddah; a 700-MW, 23-mmg/d plant at Shuqaiq in the country's southwest; and a 2,500-MW, 176-mmg/d plant at Ras Az Zour in the Eastern Province. Saudi Arabia's Saline Water Conversion Corp. (SWCC) has estimated that the country will need to spend $50 billion on water projects through 2020 in order to meet the Kingdom's rapidly growing water demand. In March 2004, Gulf News reported that Saudi Arabia planned to establish 10 IWPPs by 2016, at a total cost of $16 billion.

On October 9, 2000, Saudi Arabia approved plans for setting up a new utility company (UCO) in the twin industrial cities of Yanbu and Jubail. The company, named Marafiq, was founded by the Royal Commission, the Public Investments Fund, Saudi ARAMCO, and SABIC, with local investors also holding a stake. UCO may be privatized when it becomes profitable. In the meantime, UCO has begun several water and power projects in Yanbu and Jubail. In July 2004, Marafiq issued a request for proposals (RFP) for a $2.5 billion, 2,400-MW, 79-mmg/d (of water), gas-fired IWPP in Jubail.

Besides generation, Saudi Arabia also requires additional investment in power transmission. At present, around 20% of Saudis are not connected to the national power grid. Creating a unified national grid could require over 20,000 miles of additional power transmission lines. Currently, Saudi Arabia has around 150,000 miles of transmission lines.

COUNTRY OVERVIEW

Head of State: King Fahd ibn Abd al-Aziz al-Sa'ud
Crown Prince: Abdullah ibn Abd al-Aziz al-Sa'ud
Independence: September 23, 1932 (unification)

Population (2004E): 25.8 million, including 5.6 million foreign nationals (growing around 2.4% per year)

Location/Size: Between the Arabian Gulf and the Red Sea/865,000 square miles (about 1/4 the size of the United States)

Major Cities: Riyadh (royal capital), Jeddah (administrative capital), Mecca, Medina, Dammam, Jubayl, Buraydah

Language: Arabic

Ethnic Groups: Arab (90%), Afro-Asian (10%)

Religion: Muslim (100%) - predominantly Sunni, with a minority Shi'ite population mainly concentrated in eastern Saudi Arabia

ECONOMIC OVERVIEW

Currency: Riyal

Market Exchange Rate (12/21/04): US$1 = 3.75 riyals

Gross Domestic Product (GDP - market exchange rate) (2004E): $247.2 billion

Real GDP Growth Rate (1995-2002 average): 1.7% (2003E): 7.2% (2004E): 6.1% (2005F): 6.2%

Inflation Rate (consumer prices) (2003E): 0.6% (2004E): 0.4% (2005F): 1.5%

Unemployment Rate (2004E): 14% (unofficial estimates are higher)

Current Account Balance (2003E): $29.7 billion (2004E): $11.6 billion (2005F): $13.1 billion

Major Trading Partners (2004): Japan, United States, European Union

Merchandise Exports (2004E): $112.3 billion (mainly crude oil and petroleum products)

Merchandise Imports (2004E): $36.6 billion (mainly industrial goods, metals, food)

Merchandise Trade Balance (2004E): $75.7 billion

Oil Export Revenues (2003E): $77 billion (2004E): around $100 billion

Oil Export Revenues/Total Export Revenues (2004E): 90%-95%

Public Debt (2004E): $176 billion (note: external debt is estimated at $39 billion)

Reserves of Foreign Exchange and Gold (2004E): $23 billion (note the country has significantly more in total "foreign assets")

ENERGY OVERVIEW

Minister of Petroleum and Mineral Resources: Ali bin Ibrahim al-Naimi (since 8/95)

Minister of Water and Electricity: Abdallah al-Husayn (since 4/04)

Proven Oil Reserves (1/1/05E): 261.9 billion barrels (includes half of Divided/"Neutral" Zone)

Total Oil Production (2004E; includes NZ): 10.4 million barrels per day (bbl/d), of which 9.1 million bbl/d was crude oil, 1.2 million bbl/d was natural gas liquids (NGLs), and 80,000 bbl/d was "other liquids" (including MTBE)

Total Oil Production (2003E; includes NZ): 9.9 million barrels per day (bbl/d), of which 8.8 million bbl/d was crude oil, 1.0 million bbl/d was natural gas liquids (NGLs), and 80,000 bbl/d was "other liquids" (including MTBE)

OPEC Crude Oil Production Quota (effective 11/1/04): 8.775 million bbl/d

Crude Oil Production Capacity (12/04E): 10.5-11.0 million bbl/d

Total Oil Consumption (2004E): 1.67 million bbl/d

Net Oil Exports (2002E): 7.0 million bbl/d (2003E): 8.3 million bbl/d (2004E): 8.7 million bbl/d

Major Oil Customers (8/04E; approximate exports): United States (1.9 million bbl/d); OECD Europe (1.4 million bbl/d); Japan (1.2 million bbl/d); South Korea (838,000 bbl/d); India (345,000 bbl/d); China (310,000 bbl/d); Taiwan (210,000 bbl/d)

Crude Oil Refining Capacity (1/1/05E): 1.745 million bbl/d

Natural Gas Reserves (1/1/05E): 235.0 trillion cubic feet (Tcf) (includes half of NZ)

Natural Gas Production/Consumption (2002E): 2.0 Tcf

Electric Generating Capacity (2003E): 26.6 gigawatts (all thermal)

Net Electricity Generation (2002E): 138.2 billion kilowatthours

ENVIRONMENTAL OVERVIEW

Total Energy Consumption (2002E): 5.1 quadrillion Btu* (1.3% of world total energy consumption)

Energy-Related Carbon Dioxide Emissions (2002E): 328.8 million metric tons (1.3% of world carbon dioxide emissions)

Per Capita Energy Consumption (2002E): 218.7 million Btu (vs. U.S. value of 339.1 million Btu)

Per Capita Carbon Dioxide Emissions (2002E): 3.8 metric tons (vs. U.S. value of 5.5 metric tons of carbon dioxide)

Energy Intensity (2002E): 17,820 Btu/$ nominal -- PPP (vs U.S. value of 9,348 Btu/$ nominal)**

Carbon Dioxide Intensity (2002E): 1.1 metric tons/thousand $nominal -- PPP (vs U.S. value of 0.55 metric tons/thousand $)**

Fuel Share of Energy Consumption (2002E): Oil (59.8%), Natural Gas (41.2%)

Fuel Share of Carbon Dioxide Emissions (2002E): Oil (66.2%), Natural Gas (33.8%)

Status in Climate Change Negotiations: Non-Annex I country under the United Nations Framework Convention on Climate Change (ratified December 28th, 1994). Not a signatory to the Kyoto Protocol.

Major Environmental Issues: Desertification; depletion of underground water resources; the lack of perennial rivers or permanent water bodies has prompted the development of extensive seawater desalination facilities; coastal pollution from oil spills.

Major International Environmental Agreements: A party to Conventions on Climate Change, Desertification, Endangered Species, Hazardous Wastes, Law of the Sea and Ozone Layer Protection.

* The total energy consumption statistic includes petroleum, dry natural gas, coal, net hydro, nuclear, geothermal, solar, wind, wood and waste electric power. The renewable energy consumption statistic is based on International Energy Agency (IEA) data and includes hydropower, solar, wind, tide, geothermal, solid biomass and animal products, biomass gas and liquids, industrial and municipal wastes. Sectoral shares of energy consumption and carbon emissions are also based on IEA data.

**GDP from OECD figures based on Purchasing Power Parity (PPP) exchange rates

Oil and Gas Industries

Organization: The Supreme Petroleum Council governs the nationalized oil industry, including Saudi Arabian Oil Co. (Saudi Aramco) - crude production, refining and marketing; Saudi Basic Industries Corp. (SABIC) - petrochemicals.

Major Foreign Oil Company Involvement: AOC, BP, Eni, ExxonMobil, Occidental, Repsol, Shell, Sinopec

Major Oil Terminals: Ras Tanura (world's largest offshore oil loading facility, on the Persian Gulf; 6 million bbl/d capacity), Yanbu (on the Red Sea, fed by Petroline; 5 million bbl/d capacity), Jubail, Ras al-Ju'aymah (on the Persian Gulf northwest of Ras Tanura; 3 million bbl/d capacity), Jiddah (on Red Sea south of Yanbu), Jizan (on Persian Gulf, refined products), Ras al-Khafji (on Persian Gulf in the Saudi-Kuwaiti Divided/"Neutral" Zone, crude oil), Rabigh (on Red Sea, north of Jiddah, crude oil and refined products), Zuluf (offshore Persian Gulf, linked to Zuluf oil field)

Major Oil Fields: Abqaiq, Abu Safa, Berri, Ghawar, Khursaniya, Najd, Qatif, Safaniya, Shaybah, Zuluf (in addition, Khurais and Manifa are partially developed but mothballed as of 12/04)

Major Pipelines (capacity - million bbl/d): Petroline (4.8), IPSA (1.65 -- closed since August 1990), Tapline (0.5 -- closed since 1984), Abqaiq-Yanbu NGL line (0.3)

Major Refineries (capacity, 1/1/05E): Aramco - Rabigh 400,000 bbl/d, Ras Tanura 300,000 bbl/d, Yanbu 190,000 bbl/d, Riyadh, 120,000 bbl/d, Jeddah 60,000 bbl/d; Saudi Aramco/Mobil - Yanbu 340,000 bbl/d; Petromin/Shell - al-Jubail 305,000 bbl/d; Arabian Oil Company - Ras al-Khafji 30,000 bbl/d

Sources

Agence France Presse
Alexander's Gas and Oil Connections
APS Review Gas Market Trends
APS Review Oil Market Trends
Bloomberg
BBC Summary of World Broadcasts
Business Week
Cambridge Energy Research Associates
Chemical News and Intelligence
CIA World Factbook
Dow Jones
Economist Intelligence Unit (EIU) Business Middle East and ViewsWire
Energy Compass
Financial Times

Global Insight
Gulf News
Hart's Middle East Oil and Gas
International Market Insight Reports
LPG World
Middle East Economic Digest (MEED)
Middle East Economic Survey (MEES)
Middle East Newsfile
New York Times
Oil Daily
Oil and Gas Journal
Petroleum Economist
Petroleum Finance Company
Petroleum Intelligence Weekly
Power Engineering International
Reuters
The Times (London)
U.S. Energy Information Administration
World

INDEX

A

B

C

V

W

Y

R

S

T

U